SOCIAL INNOVATION AS POLITICAL TRANSFORMATION. THOUGHTS FOR A BETTER WORLD

Editorial design:

 Gonzalo Caceres - Studio de Création
www.caceres.be

 European Spatial
Development &
Planning Network

www.esdp-network.net

 Innovating spatial development
planning by differentiating land
ownership and governance

www.theindigoproject.be

FLANDERS
INNOVATION &
ENTREPRENEURSHIP

Flanders
State of the Art

Social Innovation as Political Transformation
Thoughts for a Better World

Edited by

Pieter Van den Broeck
Department of Architecture, KU Leuven, Belgium

Abid Mehmood
Sustainable Places Research Institute, School of Social Sciences, Cardiff University, UK

Angeliki Paidakaki
Department of Architecture, KU Leuven, Belgium

Constanza Parra
Department of Earth and Environmental Sciences, KU Leuven, Belgium

Cheltenham, UK • Northampton, MA, USA

Published by
Edward Elgar Publishing Limited
The Lypiatts
15 Lansdown Road
Cheltenham
Glos GL50 2JA
UK

Edward Elgar Publishing, Inc.
William Pratt House
9 Dewey Court
Northampton
Massachusetts 01060
USA

Paperback edition 2020

A catalogue record for this book
is available from the British Library

Library of Congress Control Number: 2019951579

This book is available electronically in the **Elgar**online
Social and Political Science subject collection
DOI 10.4337/9781788974288

ISBN   978 1 78897 427 1 (cased)
ISBN   978 1 78897 518 6 (paperback)
ISBN   978 1 78897 428 8 (eBook)

Printed and bound in the UK by CPI Colour

Dedicated to Isabel André, a good friend,
who passed away on 3 April 2017.

Isabel was a great, good and warm person,
a creative scholar, a wonderful teacher and tutor.
I enjoyed so much working with her and her
colleagues from Lisbon on the creative side of
social innovation, and the role of arts and
community (re)building.
I remember all the beautiful places in Europe and
the Americas (Canada of course) which we visited
together, where we met to work interactively,
communicate with people,
laugh about the world and ourselves,
enjoy wonderful food from the earth,
share ideas and observations ...

Frank Moulaert, 4 April 2017

TABLE OF CONTENTS

01

INTRODUCTION

1. SOCIAL INNOVATION AS AN ALTERNATIVE NARRATIVE

◆

Pieter Van den Broeck, Abid Mehmood,
Angeliki Paidakaki and Constanza Parra

THE AMBITION OF THIS BOOK

In a world threatened by several crises, including collapse of ecosystems, hegemony of market thinking over the economy, rolling back of the state, diminution of democracy to moments of election and consensus, concentration of wealth, predominance of global corporative conglomerates, rapidly changing identities and loss of socio-cultural diversity, there is an urgent need of an alternative and persuasive narrative. Although threatened (and often plagued) by various economic-centred instrumentalizations, the theory of social innovation provides such an alternative. This theory combines a critical, profound and holist understanding of societal dynamics with a focus on socially innovative practices, responding to unaddressed needs, empowering the deprived, and changing the social relations. Social innovation encompasses collective action that succeeds in changing the world for the better, by mutually embedding social, economic and ecological systems, recreating solidarity-based community relations, stimulating a fair distribution of resources, establishing regenerative economies, and supporting socio-cultural, discursive and cognitive diversity.

This book aims to introduce the work of a collective of academics on social innovation and socio-political transformation. It offers a critique to the dominance of market-based logics and extractivism in the age of 'caring neo-liberalism'. Calling for systemic change, the authors invite the reader to engage in the analysis and practice of socially innovative initiatives and, by doing so, contribute to the co-construction of a sustainable, solidarity-based and re-generative society. As such, the book intends to offer various interpretations of the interconnectedness of social innovation and socio-political transformation, which are part of a more or less coherent socio-scientific project expressed in shared publications, pedagogies, research projects, training through workshops and summer schools, exchange visits, action-research, and pro-activist practices.

THE NATURE OF THIS BOOK

This book was originally conceived, along with an international academic conference, to celebrate the academic career of Frank Moulaert as a leading scholar in social innovation. The celebratory pitch was later extended to the larger working of the ESDP Network, a group of scholars of all ages and disciplines from across the world, active since the 1980s (see e.g. http://esdp-network.net/). To start with, authors were invited to write about specific aspects of Moulaert's conceptual underpinnings and ESDP Network's empirical work in the wider intellectual thinking, through different forms of expression.

The book has subsequently transpired as a socio-political academic statement endorsed by renowned as well as emerging academics with clear ideas on how to change the world. It is also an introduction to the work of a generation of scholars, keen on social innovation and socio-political transformation, a manual for the engaged researcher, and an invitation to nuanced thinking and acting in the quest for a more egalitarian society. The book encompasses a wide range of topics covered by experts in the field into a collection of research and analyses independent of the conventional academic norms and constraints.

This book expresses a belief in the power of critical reasoning, the need to root collective action in high quality analysis, and the will to connect policy making to research agendas. The analysis is critical, questioning inequalities, exclusionary mechanisms and selectivities, interacting with power relations and power structures. The authors argue for an integrated approach, and a holist understanding and interdisciplinary research in all domains of research and action, assuming that a complex world needs sophisticated understandings. This ambition relates to a deep interest in various - including non-academic - modes of knowledge production and research methodology, covering a

© Dimitra Vinatsella

dynamic/ontogenetic perspective on ontology and a critical-realist perspective on epistemology. According to the authors, knowledge is socially constructed. This requires awareness of and reflection on the interaction between researchers and other actors, entailing the need for co-constructed research questions, shared problematizing and theory building, a transdisciplinary approach and careful and engaged action research. Of course, social innovation, as one of the key phenomena and concepts in the work of a majority of the authors, is widely discussed in the book, including the concept's different dimensions, its main underlying research questions and the critique on how the concept has recently been instrumentalized to serve market fundamentalism. The study of how particular social innovations resist the destructive forces of the market and add up to a broader socio-political transformation belongs to the core of the authors' research and action interests, and shows how both strategic action and structural mechanisms need to be understood in their dialectical interactions. This implies a multi-scalar and relational perspective on space and place-making, as applied in numerous analyses of socially innovative case studies.

In sum, this book is a statement, including not only a critique of the age of neo-liberalism challenging the dominance of market-based thinking, but also a call for systemic change. The authors express a worldview that is shared by a large network of academics some of whom have been collaborating for the past thirty years. As such, the book is not only an inspiration for academia in many ways, but also for social movements and their protagonists challenging the dominance of the powers that be and changing the world, and for politicians who want to appreciate the contemporary ways of thinking and get inspiration on how to better attend the needs of the communities they serve.

Finally, this book is a manual for the engaged researcher, which is evident from the critical questions addressed, the critical-realist epistemological approach, the heterodox

© Els Dietvorst

theories that are mobilized, the will to bridge science and arts, the social and the ecological anatomy and action, the attention to solidarity and reciprocity in human relations, the analysis of the needs of all actors, and the empowerment of the underprivileged. Many chapters in the book echo the researchers' concerns about poverty, exclusion, inequality, competition, resource grabbing, the growth fetish, and attempts to forward the market as the main organizing mechanism of our society and the ecosystems it interacts with. As such, the authors invite you to join this endeavour to analyse, reflect on, and act in the world we live in today to co-construct a solidarity-based and regenerative alternative.

ORGANIZATION OF THE BOOK

To build a consistent, though diverse, overview on social innovation and socio-political transformation, the editors drafted a list of key topics in this field. To do justice to the diversity of the authors' interests, these topics were to be addressed in three different ways - essays, text-boxes and stories - after which particular authors were contacted to offer their contributions. The response was overwhelming. Once all texts were made available, the book was reorganized into 15 thematic sections, showing the variety, complexity and richness of the authors' thinking and practices.

As such, 55 authors contributed to a variety of essays, short explanations and illustrative cases, covering work on: Keynesian economics and regional development; migration; integrated area development; poverty; territorial innovation systems; heterodox political economy; urban regeneration and the new urban policy; governance, social innovation and solidarity; neighbourhoods and communities; culture, arts and social inclusion; spatial development planning; the sociology of knowledge, science-society relations and pedagogy; social innovation, sustainability and socio-ecological systems; underdevelopment and global North-South relations. Most of these contributions combine historical insights, current debates and avenues for the future.

Inevitably, a few topics may be missing. The editors are however confident that the extensive bibliography will enable readers to explore additional aspects of social innovation and socio-political transformation for themselves.

ACKNOWLEDGEMENTS

Since this book is a collective effort, many people need to be thanked. The editors are grateful to all authors for their spontaneous positive response, high quality contributions to the book and willingness to tune into a concept which might not have been totally clear from the beginning.

Thanks and apologies are due to all who could have been here but were not included due to limited space, and to those who tried but could not contribute due to various reasons. Special thanks go to Diana MacCallum for helping with a 'light-touch' language check and a 'proper' bibliography. Thanks to Ruth Segers & co. for providing the cover picture of this book. Special thanks to Maarten Loopmans for making tailored stimulating sketches and to Els Dietvorst for her wonderful drawings. Last but not least, we are grateful to Flavia Martinelli and Erik Swyngedouw for supporting the project all along.

The publication of this book was possible thanks to the support of the Flemish Agency for Innovation and Entrepreneurship (VLAIO) and their programme SBO for the research project INDIGO by KU Leuven, UAntwerp, Harokopio University and OMGEVING cvba (see www.theindigoproject.be).

Many of the authors wouldn't have been able to spend lifelong careers in critical research if there hadn't been people supporting them. So, the editors, and surely also all authors, thank all those families, friends and relatives bearing with the efforts and time that academic careers unfortunately demand.

Finally, the editors wish to dedicate this book to Isabel André, a good friend of many of the people involved in this project, who sadly passed away on 3 April 2017, during the making of this book. Her draft chapter may be one of the last pieces she wrote, which reflects her passion and commitment. Thank you, Isabel, for everything you have given us. We will continue your work and extend your gentleness.◀

Petrified forest, Lesvos © Constanza Parra

02

Toulouse © Constanza Parra

FROM KEYNESIAN ECONOMICS TO SOCIAL INNOVATION

2. ANIMATING SOCIAL CHANGE: POLITICAL TRANSFORMATION AND/OR SOCIAL INNOVATION?

◆

Erik Swyngedouw

This chapter is a brief reflection of process and forces of social and political transformation over the past few decades, and how my own inspirations and work have evolved, sometimes from the sidelines but often through direct participation with the theoretical and empirical underpinnings advocated by Frank Moulaert. These include social and economic dynamics of capitalism with a particular focus on the mechanisms of disempowerment and growing inequality and with a critical eye towards identifying and nurturing strategies for emancipatory transformation through socially innovative practices. For about thirty years, my intellectual work and Moulaert's academic trajectory have unfolded in close conversation, occasional joint writing and recurrent, albeit always comradely, intense debate. Our joint perspective nonetheless was firmly focused on transforming social and political-economic relations in the direction of a more equal, free and inclusive social order, one that would permit each and every one to thrive and nurture their lives and that of those close to them in an emancipatory and fulfilling manner.

From the very early days in the late 1970s onwards, Moulaert's academic work revolved centrally around the institutional arrangements and social support structures that galvanize innovative social change in a broadly socialist or strong social-democratic manner. Trained as an economist and steeped in post-Keynesian economic thought, his early work with POLEKAR – an influential progressive left-of-centre Flemish think-thank that brought together radical social scientists – excelled in developing and proposing progressive social and economic policies in the hope and expectation to revive a socialist political programme and activism in the midst of the early signs of a profound crisis of both capitalism and social-democracy (POLEKAR, 1981, 1985). This post-68 period still held high the possibility of a close conversation between left intellectuals and progressive political parties.

In particular, Moulaert's work on the social and economic role of immigrants and the deconstruction of the blatant – although with hindsight rather mild (given the present mainstreaming of racist narratives) – discourse of the social profiteering and state-dependency of immigrant groups had a profound influence. This work marked already what would be one of the enduring convictions animating his subsequent work, namely his unwavering belief in the power of reason and the political importance of a verifiable truth as a necessary path to sane and socially sensible policies. In this sense, Moulaert has always been a modernist for whom the power of clear thought, hard facts and verifiable truths would pave the way to a more humane society. Post-truth politics were not yet invented and would never figure in his vocabulary. With the razor-sharp analytical clarity that would mark all his scholarly work, Moulaert demonstrated the utter nonsense and factual fallacies on which much of the then racist discourse rested. His co-authored book – Racisten hebben Ongelijk (Racists are Wrong) – unequivocally demonstrated the myth of immigrants as social security dependent profiteers (Bollen and Moulaert, 1983).

Our subsequent joint work and interest led us to consider more centrally the role and dynamics of institutions, formal and informal social regulations, and social practices in choreographing the medium-term dynamics of capitalism. The French *Regulation School* became a major inspiration here, enhanced with institutional post-Keynesian perspectives from Moulaert's side and more Marxist leanings from my part (Moulaert and Swyngedouw, 1989). While my work initially focused on socio-technical transformations in the transition from Fordist to flexible accumulation, Moulaert focused on the tremendous economic and institutional transformations associated with the rising prominence of the service sector and associated labour market dynamics (Moulaert, 1987; Moulaert and Daniels, 1991; Cooke, et al., 1992). Questions of social conflict and political struggle remained paramount in excavating the rapid and innovative transformation of institutional and socio-economic forms. Indeed, one of the major contributions throughout Moulaert's academic career focused on extending the notion of *innovation* from the purely technical and "scientific" to include and foreground the foundational role of social innovation in initiating and nurturing institutional and political transformation. He contends that social innovation and the conflict-ing process of its institutionalization is the necessary basis for emancipatory transformation and inclusive development (Moulaert, 2000). Throughout his writings, he sought out, explored, examined, and supported new and innovative social practices that pointed at a horizon beyond the existing and opened concrete and realistic perspectives from a more inclusive social order. For him, social innovation offered the sparkle and the experimental foundation for socially progressive transformations and for nurturing the search for a more humane, socially equitable and ecologically sane social order (Moulaert and Sekia, 2003).

In the process, Moulaert's work always oscillated between the critical analysis of the reactionary forces and actors that mitigate against change and pursue strategies to make sure that nothing would really change on the one hand and the scholarly defence of the agencies and possibilities of the subaltern in their active search for empowering alternatives on the other. The quilting points that provided the nexus through which this Machinean analysis was pursued revolved around *poverty, exclusion,* and *disempowerment*. Skilfully meandering between critiquing forms of *innovation* that just re-enforced the neoliberal status quo and exploring forms of social innovation that contained possibilities for progressive transformation, his work invariably sided with the progressive potential lurking within the interstices of the existing order.

These themes were systematically explored in a series of successive large European research projects, each of which combined in-depth empirical research with advanced critical theoretical work that combined new institutionalism with post-Keynesian and Marxist political economy. With extraordinary determination and not a little bit of diplomatic negotiation, these projects succeeded in achieving successfully the usually impossible compromise between the policy-oriented and policy-relevant research the European Commission customarily demanded on the one hand and advancing critical urban theory and practice on the other.

URSPIC, SINGOCOM, DEMOLOGOS, and SOCIAL POLIS are just some of the projects that both of us were involved in. This sequence of projects also charts a theoretical and political trajectory over twenty years of collaborative research. While URSPIC was largely diagnostic, tracing the geographically constituted processes of budding neoliberalization as it became institutionalized through the urban process and etched into the socio-economic and physical fabric of the city, the subsequent programmes focused more directly on the dynamics of progressive political-economic transformation. URSPIC was a project that continued to be embedded in a critical scholarly practice of excavating the really existing socially uneven and class-based character of neoliberalizing urbanization (Moulaert et al., 2003). This analysis was still infused with the dominant view that substantive critical analysis, which would speak the "truth" of the situation, might galvanize progressive political reaction and resistance.

Over the subsequent few years and with a deepening and widening entrenchment of neoliberalization, it became nonetheless increasingly evident that the assumption of the performative political role of critical theory and empirical analysis could not be sustained much longer. Indeed, despite radical critique of neoliberalization becoming dominant in critical academic work, the actual practices of neoliberal transformation in the real world became truly hegemonic. While the post-68 view insisted that critical analysis would support and articulate with progressive political tactics, the late 1990s experiences indicated that there might not be a clear and identifiable relationship between critical thought and research on the one hand and progressive political action on the other. The subsequent research projects, therefore, moved much more decidedly in the direction of discerning radical social and political possibilities and practices that were already germinating within the interstices of actually existing neoliberalization. These socially innovative practices signalled both a growing discontent from within civil society with the state of the situation on the one hand and a proliferating experimentation with new and alternative modes of life and socio-economic or socio-ecological interaction on the other (Moulaert et al., 2005, 2010, 2013b; MacCallum et al., 2009).

It is precisely these innovative, progressive, if not radical, forms of socio-spatial experimentation that would be the leitmotiv of much of Moulaert's subsequent leading-edge research, extending his concerns from solidarity-based forms of social interaction to include socio-ecological matters as well (Parra and Moulaert, 2016). A subtle shift occurred in these successive projects, one that moved the gaze from a predominantly analytical perspective to a perspective that foregrounded the political possibilities already embedded and prefigured by the combination of innovative experimentation with alternative practices in a wide range of both urban and rural settings. Indeed, in a context of near-hegemonic closure of radical alternatives by the elite-led post-democratic techno-managerialist approach and the colonization of everyday life by neoliberal doctrines, the possibilities of opening new political avenues seemed foreclosed (Wilson and Swyngedouw, 2014). It is precisely in such autocratic political environment that a wide range of new and often innovative solidarity-based forms of social and community experimentation proliferated. In short, the first decade of the twenty-first century signalled both a closure of alternative mainstream policies and new possible openings pioneered by new forms of subjectivation. The politics of resistance to class-based neoliberalization, which allegedly became the horizon of the possible in the great late 1990s anti-globalization protests – from Seattle to Genoa and beyond – morphed into something radically different, and this required urgent thought, theoretical articulation and empirical verification. The real possibilities for a different world did not seem to reside any longer in demonstrating the " truth" by enlightened critical scholars, accompanied with impotent calls to change the infernal course of social disintegration ushered in by rampant neoliberalization, but rather in really existing new experimental socio-ecological practices, allied with new forms of activist scholarship.

Indeed, all manner of people began to embark on a different trajectory, rupturing the routines of impotent protest, and to become politicizing subjects through engaging in, organizing, and practising new forms of life-in-common, while undermining or at least side-lining the vestiges of the established order. The exploration of these variegated innovative socio-spatial and socio-ecological practices became the key focus of Moulaert's academic work, ushered forward with a theoretical project to construct a robust theory of social innovation that is intimately related to practices of social emancipation, solidarity and democracy. While critically returning to institutional analysis and the analysis of collective social practices from the past, these insights were imaginatively re-interpreted to articulate with the emergence of new social movements and strategies. The nurturing of an agonistic space where alternative and oppositional practices could thrive became one of the key objectives of this research programme. Invariably, it turned the academic scholar into a scholar-activist, something that had marked Moulaert's academic and political life from the beginning, but now began to move to centre stage as a symbiotic activity whereby theory infused social activity and subjectivation, and vice versa. His innate activist desire now fused seamlessly with his intellectual agenda.

A few key insights emerge from this trajectory. First, the focus squarely resided in the prioritization of emergent social practices as opening up the concrete possibilities for social and political change. Second, the reversal of the axis of time. Whereas earlier work pointed at possible futures, but ones conditional on the correct organization and political processes to nurture this progressive dawn to come, his work gradually morphed into an ontological position that discerns the real possible future lurking in the realities of already existing practices. In other words, future possibilities are already brewing within the interstices of the existing configuration. Third, his intellectual gaze is increasingly focused on privileging processes of subjectivation, that is the transformation of being-as-object-of-law to the becoming-subject, that is the active process whereby individuals or social groups no longer just live life, but begin to make life, that is to actively take control of, organize, and govern their lives in common. Fourth, this process of social transformation points indeed to the transformative capacity of innovative social practices subtracted from the state, that is unfolding at a critical distance from the state (at whatever scale) without, however, following into a naïve or romantic anarchism. Emancipatory transformation takes place not through, but despite, the state. Nonetheless, the institutionalizing power of the state is readily acknowledged and supported. Fifth, this leads to the confirmation that socio-spatial transformation resides in the impregnation of innovative social practices that point to the inauguration of more democratic and egalitarian modes of life with new forms of politicization. The latter implies the process of rendering universal (in the face of often formidable opposition) the practices already operating in these embryonic but interstitial spaces of social innovation.

Here precisely resides the convergence between Moulaert's current scholarly work and my own recent attempts to re-think the possibilities of emancipatory and progressive socio-ecological transformation at the beginning of the twenty-first century. While much progressive thought still dwells in the humanitarian-liberal view of the need to foster sustainability, adaptation, resilience and management to the (class) givens of the situation and the social management of the excesses of capital, our work consistently points at the possibilities and necessities of emancipatory transformation beyond the existing as they manifest themselves in the present socio-political conjuncture. Moulaert prioritizes – and quite correctly to my mind – the pivotal importance of socially innovative and experimental practices of self-organization and common governance and of the heterogeneous and often locally specific forms of struggle and conflict accompanying such experimental socio-ecological labs. My work, in contrast, tends to prioritize *the political* moment and process. The latter is concerned with the organizational forms, strategies, and practices that attempt to universalize – or spatialize – these locally embedded innovative socio-ecological modes of life, and with the organizational forms through which such new forms of politicization may unfold. Such politicizing process requires a particular set of skills, terrains, and forms of subjectivation. It is the question of how locally specific forms of social transformation become universalized, often in the face of radical opposition by conservative forces. Socio-ecological transformation of a progressive kind clearly requires both social innovation and sustained processes of politicization unfolding in close relation and affinity.

Ultimately, Moulaert's work and that of many of his fellow travellers on his academic journey are infused with the conviction that the twenty-first century world requires more than adaptation to and massaging of existing neoliberal globalization and its associated twin process of accentuated socio-spatial inequality and uneven environmental degradation. A progressive transformation is therefore not only desirable, it is an utmost necessity if the practices of equality, freedom, solidarity and socio-ecological common sense are to be realized. His intellectual legacy will remain a vital foundation on this slippery road to a better world. Indeed, there is no other alternative.◄

03

MIGRATION AND INTEGRATED AREA DEVELOPMENT

3. THE METAMORPHOSES OF A RESEARCH FIELD: FOREIGN LABOUR MIGRATION IN BELGIUM 1945-2016

Albert Martens

After the Second World War, Belgium repeatedly relied on workers from abroad. As early as 1944-1945, Belgium set no fewer than 60 000 German prisoners of war to work in the coal production industry as part of the "national industrial reconstruction" effort. In 1945-1946, they were succeeded by 70 000 Italians. Following a tragic mining disaster in Marcinelle on 8 August 1956, which cost the lives of more than 270 miners, workers were recruited from Spain and Greece. Later, during the "golden sixties" era (1962-1972), thousands of work permits were issued to workers from Turkey and Morocco (Martens, 1973). As in France, and later also in Germany and the Netherlands, the unskilled, low-paid and often unhealthy and dangerous work was assigned to foreigners (aliens). The recruitment system used in the mining industry later spread to other economic sectors including the steel, textile, chemical and construction industries, as well as the services sector (transport, distribution, cleaning, hospitality industry).

The spread and generalization of these labour market developments did not go unnoticed by the social scientists. Phenomena such as ethnic stratification of the labour market, segregation of jobs and professions, racism and discrimination all came to the surface, clearly and visibly. New tensions in the traditional labour relations divided workers. This also presented a challenge to the trade union organizations. The "coloured" character of the working-class districts in Belgian towns and cities had a knock-on effect on consumption, education, health, culture, and so on. Persistent pressure from (extreme right-wing) anti-immigration and racist political factions meant that by the end of the 1970s, "migrant policy" was a topic that could no longer be ignored.

This was the historical context which led to the "migrant issue" being placed on the research agenda in Flanders.

THE FORGING OF AN ALLIANCE BETWEEN SCIENTISTS AND THE COLLECTIVE CALL OF RESEARCHERS: A FIRST ATTEMPT TO DEMARCATE THE RESEARCH FIELD (1974-1985)

After 1974, the economic recession brought tensions in the Belgian system: not just on the social front, but also in the financial, economic, geographical and political arenas (Moulaert and Martens, 1988). Several social scientists became increasingly aware of the unequal position and treatment of the foreign workers' population – not just at work, but also in areas such as housing, health, their children's education, and so on (Martens and Moulaert, 1978, 1981).

The upshot was that on 25-26 May 1984 a colloquium was organized to discuss the situation of foreign minorities in Belgian Flanders (*De toestand van de buitenlandse minderheden in Vlaanderen-België*) (Moulaert and Martens, 1985, 1986). The colloquium was the result of a unique alliance between leading scientists from a variety of disciplines and universities, the Belgian National Fund for Scientific Research (NFWO), the Flemish Government, the Flemish Consultative Committee for Migration and the universities of Brussels and Leuven, which a year later published an *Integration Policy Report* (Rapport Integratiebeleid) (Hobin and Moulaert, 1986).

During the colloquium, a common diagnosis of the situation was formulated and shared by various researchers:

> The long term and permanent residence of foreigners (more than 40 years, spending two or three generations) is increasingly leading to discontent and tensions. Long term cohabitation has meant that mutual comparison and weighing of the respective pros and cons has come into ever sharper focus. (...) The awareness of 'relative inequality' is increasing and putting pressure on social research, which will increasingly be more systematically focused on the '(un)equal treatment' of foreign minorities. Evidence of the various forms of discrimination and discriminatory practices is growing day by day and fuels the debate on the experience, admissibility, condemning or combating of these forms and practices. (Moulaert and Martens, 1985).

Worth noting here is the clear perception of the insurmountable contradictions and paradoxes for the "Belgian" society with the immigrants at that time, such as:

- The redefining of basic democratic principles such as freedom, equality and solidarity; are those values reserved solely for "natives", or are they also granted to others (non-natives)? If so, how? And in what form?

- The confrontation between different legal cultures; nationality law or domicile law in building legal certainty within the legal system.

- The commitment to a positive integration policy together with a strict restrictive/repressive admissions policy (immigration freeze) for new immigrants.

- The explicit questioning of the rights of aliens (including the right of residence) and the admission and acceptance of (right-wing) opinions and views on the need to repatriate foreigners, including those who have settled in the host country.

- The policy model to be followed: a spatial policy such as geographical/territorial distribution; mixing of the different population groups; or a specific policy focus within a general disadvantage policy (Martens and Moulaert, 1985: 25-7).

At the end of this conference, many scholars and researchers from the Flemish universities (Antwerp, Brussels, Ghent, Hasselt, Leuven) agreed to join the *Integration Policy Study Committee* and to work together to solve some of the contradictions.

THE INTEGRATION POLICY STUDY COMMITTEE: ON THE BRIDGE OF A SCIENTIFIC FLAGSHIP (1985-1989)

The Study Committee was given the following remit:

- Drawing up a profile of people and society that would serve as a philosophical basis for the integration policy.

- Describing the content of a positive integration policy for the different societal subsystems.

- Compiling an inventory of the position of foreign minorities in Belgium and abroad.

- Critically evaluating the existing integration policy in Belgian Flanders and in a number of neighbouring countries.

- Drawing up a programme of priorities for the integration policy in the short and medium term.

- Working up legal, budgetary and practical proposals for putting the priorities programme into practice.

The programme resulted in an intensive sharing of ideas and collaboration between research teams from a variety of disciplines, universities and research centres in Antwerp, Brussels, Hasselt, Leuven, Ghent, but also Liège and Mons. On 22 February 1989, the NFWO contact group for *Scientific Research on Foreign Minorities (Wetenschappelijk Onderzoek Buitenlandse Minderheden)* published a memorandum outlining the establishment of coordinated research for contribution to adequate decision-making and policy, and a broad and robust "science" for a better society (Hubeau et al., 1989). A programme of research proposals was drawn up covering thirteen research fields (political participation, municipal policy, societal problems, demographics, legal position, education, health, employment, geographical localization, housing, welfare, sport and leisure, media), for both survey and fundamental research. The proposals included realistic timelines and financial projections.

WHAT RESEARCHERS SOW, THE GOVERNMENT REAPS (1989-2000)

Different circumstances, including the rise of anti-immigration political parties, openly racist standpoints and discrimination, forced the Belgian government to appoint a Royal Commissioner for Migration Policy on 7 March 1989. The Commissioner was charged with 'researching and proposing the measures needed to address the migration issue' (Article 2). It explicitly identified employment, housing and education as key areas.

The researchers, through the Integration Policy Study Committee, ultimately achieved their goal: the "migration issue" was placed on the political agenda at the highest policy level, that of the Prime Minister. Numerous studies were conducted, and reports published in the ensuing years. The Royal Commissioner de facto assumed the role of the former Study Committee and sought to translate the recommendations of the researchers into policy decisions (Brans et al., 2004).

THE RE-ORIENTATION OF THE RESEARCH TOWARDS RACISM AND DISCRIMINATION AGAINST MIGRANTS AND ETHNIC MINORITIES (2000-2016)

Statistical research on specific population groups (in this case immigrants, aliens or ethnic minorities) assumes that specific and uniform characteristics can be assigned to each population group. And this was the case until the early 1980s. The division of the population into two categories – Belgian/foreigner – caused little confusion: a person who held Belgian nationality was a (native) Belgian, and had (usually) been born of Belgian parents. Someone who was an alien was not and had not. Moreover, aliens could also be identified based on their nationality.

In the early 1980s, and more especially since 1985 and the ensuing years, obtaining Belgian nationality was made quicker and easier. This made it possible for the hundreds of thousands of immigrants to obtain or acquire Belgian nationality. In 2006, for example, there were more than 1.6 million people living in Belgium who had been born abroad, and more than 45% of them had obtained Belgian citizenship (Perrin and Schoonvaere, 2008: 145). However, having become "Belgians" did not mean that they were treated as such. For these "new" Belgians, often referred to as "foreign Belgians", racism and discrimination remained a part of everyday life. In order to be able to observe and establish this "statistically", it was necessary to subdivide the category "Belgian" into different subcategories based on *origin* and *migration background*. To achieve this using anonymized administrative databases, all kinds of difficulties – legal, administrative and statistical – had to be overcome. After more than ten years of effort (2000-2012), this was finally achieved (Martens, 2012).

The distribution of the total Belgian population in 2012, by origin and migration background, is shown in the following tables.

◆

DISTRIBUTION OF THE POPULATION AGED 18-60 YEARS IN 2012, BY ORIGIN

POPULATION AGED 18 TO 60 - 6,269,742

63.3% BELGIAN ORIGIN [1] 4,048,393	FOREIGN ORIGIN [2]	29.3% 1,874,076	ORIGIN UNKNOWN	7.4% 474,225
	EU-14 [3]	817,160 12.8%	Belgian: Belgian-born, one Belgian-born parent, other parent unknown 249,005	3.9%
	EU-12 [4]	159,240 2.5%		
	EU candidate states	137,205 2.1%		
	Other European	112,641 1.8%		
	North African	303,097 4.7%		
	Other African	141,019 2.2%	Belgian: Belgian-born, parents unknown 223,137	3.5%
	Near/Middle East [5]	37,748 0.6%		
	Oceania/Far East [6]	38,963 0.6%		
	Other Asian	59,104 0.9%		
	North American	13,641 0.2%		
	South/Central American	37,214 0.6%	Others 83	0.0%
	Unknown foreign origin	17,044 0.3%		

Source: Socio-economische monitoring (Interfederaal gelijkekansencentrum, 2015:15)

1 Belgian origin: people of Belgian nationality, Belgian-born and with Belgian-born parents.
2 Foreign origin: people of other than Belgian nationality or born with other than Belgian nationality, or one of whose parents was born with foreign nationality or has foreign nationality.
3 EU-14: France, Germany, Italy, Netherlands, Luxembourg, Ireland, United Kingdom, Denmark, Greece, Spain, Portugal, Finland, Sweden and Austria.
4 EU-12: Czech Republic, Estonia, Cyprus, Latvia, Lithuania, Hungary, Malta, Poland, Slovenia, Slovakia, Bulgaria and Romania.
5 Near/Middle East: Iran, Israel, Palestinian Territories, Jordan, Iraq, Syria, Lebanon, Saudi Arabia, Yemen, Oman, United Arab Emirates, Qatar, Bahrain, Kuwait, Egypt, Pakistan and Afghanistan.
6 Oceania/Far East: China, India, South Korea, Japan, Taiwan, Oceania.

DISTRIBUTION OF THE POPULATION AGED 18-60 YEARS IN 2012, BY MIGRATION BACKGROUND

POPULATION AGED 18 TO 60 - 6,394,694

BELGIAN NATIONALITY 5,536,783 — 86.6%

3rd generation — 63.3% — 4,048,393

Category	Value	%
Belgian: Belgian-born, Belgian-born parents, Belgian-born grandparents	1,201,497	18.8 %
Belgian: Belgian-born, Belgian-born parents, at least 1 EU grandparent	72,491	1.1 %
Belgian: Belgian-born, Belgian-born parents, at least 1 non-EU grandparent	6,945	0.1 %
Belgian: Belgian-born, known grandparents Belgian-born	1,493,485	23.4 %
Belgian: Belgian-born, Belgian-born parents, all 4 grandparents unknown	1,273,975	19.9 %

2nd generation — 17.0% — 1,874,076

Category	Value	%
Belgian: Belgian-born, Belgian parents, at least 1 EU-born parent	176,925	2.88 %
Belgian: Belgian-born, Belgian parents, at least 1 non-EU-born parent	149,225	2.3 %
Belgian: Belgian-born, at least 1 parent with EU nationality	202,941	3.2 %
Belgian: Belgian-born, at least 1 parent with non-EU nationality	71,933	1.1 %
Belgian: Belgian-born, parents unknown	485,844	7.6 %

Belgian nationality acquired — 6.3% — 401,522

Category	Value	%
EU	62,844	1.0 %
More than 5 yrs ago	38,294	0.6 %
5 yrs ago or less	23,920	0.4 %
Non-EU	338,678	5.3 %
More than 5 yrs ago	159,987	2.5 %
5 yrs ago or less	178,691	2.8 %

FOREIGN NAT. 857,781 — 13.4%

Entered in the Population Register — 13.4% — 857,781

Category	Value	%
EU	537,783	8.4 %
More than 5 yrs ago	258,841	4.0 %
5 yrs ago or less	278,942	4.4 %
Non-EU	319,998	5.0 %
More than 5 yrs ago	80,500	1.3 %
5 yrs ago or less	239,498	3.7 %

Belgians: Belgian-born, Belgian-born parents

Belgians: Belgian-born, Belgian EU-born/non-EU-born parent(s), or parent(s) of EU/non-EU nationality

Foreigners who have become Belgian

Non-Belgians

Source: Socio-economische monitoring (Interfederaal gelijkekansencentrum, 2015:21)

A statistical comparison between these subcategories is now possible. This has been carried out, for example, for a series of employment items (such as activity rate, unemployment rate and duration, wages, working hours, sector of employment, and so on) (Interfederaal Gelijkekansencentrum, 2015). The inferior position of employees of non-EU origin compared with EU subjects, and especially with "native Belgians" was established irrefutably.

CONCLUSION: SOCIAL INNOVATION AND THE AVAILABILITY OF VALID AND RELIABLE DATA

Following the conception of social innovation as contextual and path-dependent with focus on inclusion of excluded individuals and groups in different walks of life at different spatial scales, it is argued that social innovation relates to social justice as an ethical position (Moulaert, 2000; Moulaert et al., 2010).

But who are these *excluded groups* and *individuals*? How can they be identified in a more or less unambiguous way? Many studies underline the fact that people who are excluded suffer disadvantage and discrimination on several fronts. However, we would call for the most accurate possible description of the population, especially where we wish to study discrimination and disadvantage statistically, test possible correlations and observe trends over longer periods.

Moreover, certain characteristics may be pertinent at a certain moment in time but lose their relevance after a few years. This was clearly the case for the *foreign workers*. In order to map out the extent of their disadvantage and discrimination, it was necessary to mobilize several research teams, including lawyers, demographers and statisticians as well as economists and sociologists; and both university researchers and civil servants.

But it is still the case that the individuals and groups in question are "labelled" differently as time passes. The same categories are no longer defined (and treated unequally) as aliens, foreigners, foreign Belgians originating from an EU member state or non-EU country, but for example as Muslims. If this distinction ultimately comes to dominate, researchers will once again have to think of relevant criteria.

Observing and helping to shape social innovation implies a never-ending quest for meaningful and pertinent observations that expose these social changes clearly.◀

4. COMBATING POVERTY AND LOCAL INTEGRATED AREA DEVELOPMENT

Pavlos-Marinos Delladetsimas

Apioneering contribution to the European theoretical and policy agenda has been the development of the notion of *Local Integrated Area Development* and the systematic attempt to place it within applied policy objectives to combating poverty in European localities. The notion of Integrated Area Development refers to both local dynamics and the global realism expressed by the grassroots and the institutionalist economic approach (Moulaert et al., 1990; Moulaert, 1992). The meaning of *integrated* implies forms of cooperation (networks, partnerships, cooperative structures) and combination of local resources with structure development agents and their respective actions. These forms, however, are not necessarily restricted to local agents, assets and local movements, as the traditional grassroots approaches may require. As a matter of fact, it has been demonstrated that local development *success stories* are mostly embedded in a broader setting of cooperation (Moulaert et al., 1994). As such, the notion of local solidarity has been enriched by a regulationist interpretation of economic and social structures in which a locality operates. Hence, the term *integrated* refers to the outcome of a combination of local cooperative forms within a structure of institutional dynamics which stretch beyond the local dimensions. Similarly, the notion of *development* is seen as a combination of two approaches: the grassroots approach stipulating community rooted objectives, and the institutionalist approach seeking to weigh the economic and development potentials of a locality. Out of this, the concept of *developmental potential* is derived, which is community rooted and at the same time considers broader dynamics (social, economic, political, socio-cultural). Finally, *area* introduces a spatial dimension in this development concept. In this last component, the difference between the two approaches (grassroots and institutionalist) is less distinct. As a matter of fact, when empirical arguments on local development emerge, the appropriate spatial scale is often undermined. In most cases, administrative criteria are adopted and thus the problematic evolves around municipalities, districts, provinces, municipal unions and administrative regions (Moulaert and Swyngedouw, 1991).

The overall Local Integrated Area Development approach predominately stepped out – and was further elaborated theoretically and empirically – from the research project on *Local Development Strategies in Economically Disintegrated Areas: A Proactive Strategy against Poverty in the European Community* funded by European Commission, DG V under its Poverty III programme in 1991-1994. The project studies 29 localities in the European Union including: deprived communities (Ostiglia IT, Spatha GR, Comarca Montes de Oca ESP, Arganil PT); rural communities with light industry heritage (Maniago IT, Vigevano IT, Almeida PT); rural communities with metal and textile mills heritage (Urbania IT, Roanne FR, Castres-Mazamet FR, Agueda PT); coal mining and metallurgy communities and harbours (Rhondda Wales UK, Valenciennes FR, Charleroi BE, Dortmund GER, South Cardiff Wales UK, Lavrion GR, Elgoibar ESP, Barakaldo ESP, Calais FR, Bremen GER, Hamburg GER, Antwerp BE, Rostock GER, Liverpool UK); and special cases (Fishguard Wales UK, Sykies GR, Perama GR, Girona ESP) (See table below). The key concern was the analysis of the socio-economic, socio-political and socio-cultural structure and agency of economically disintegrated areas or localities: to identify their key socio-economic problems and their vulnerable and excluded groups; to acquire insight into the main local development strategies; and to develop the potential to establish or consolidate development strategies which could meet the needs of the communities in question. The work consisted of transversal analysis of local socio-economic development experiences, consequences for different segments of the population in the localities, and development practices especially those inhibiting a strategic policy outlook. The analysis was structured through a parallel dialectical process of theory building that started with a theoretical appraisal of the different interpretations of the role of the state (national, regional, local and supranational state) in local development. This provided at the very early stage the key variables and key issues and mechanisms to be considered regarding the role of public in relation to private agents in development strategies at the local level.

Using the variables and key issues on the role of the state, emphasis was subsequently shifted to the main features of and differences between the institutional frameworks in the various states to understand the potential effectiveness of local economic policy recipes. The transversal reading of the institutional system in the different countries, and the laws, regulations and agencies that are involved in local economic development, provided insight for further research concerning the improvement of the regional and local institutional systems to imbed, frame or steer local development endeavours. The process then embarked into the development of a double value function. On the one hand, a more positive, more systematic approach was mobilized with respect to the different variables that have been used in a transversal reading. The use of these variables provided a typology of development experiences, socio-economic problems and poverty-struck groups which had undergone the consequences of local disintegration. Some variables were also used to identify target groups which would be eligible for special measures and special policy mixtures within the overall approach to local development strategies. Also, variables related to local economic development strategies and policies – deriving from the transversal analysis – were used to structure the analysis of localities. The analysis of policy and strategy experiences stipulated the formulation of a typology of localities from a strategy and policy point of view. On the other hand, the links between the different types of development experiences and potential local innovation strategies were identified. These were the early insights towards the determination of the social innovation approach which followed in the next two decades.

In this respect, the research constituted a provocative challenge to the orthodox approaches to a social and economic innovation. In a way, it superseded the technological and technocratic view on innovation. Innovation aspects were approached by taking into account the distinct social dynamics of each locality within the broader regional settings. These dynamics could be improved and geared towards very specific goals, by means of specific proactive strategy packages for combating poverty. By definition, all these implied – as a pre-condition – a most detailed understanding of the local settings and their different socio-economic trajectories. As a result, a subtler and politically useful classification of innovation strategies for local economic development was constructed.

By reconsidering the line of reasoning developed by this pioneering research some critical observations have been reached. The project above all visualized the systematic need to combine local development strategies with strong economic targets: targets that must include the enrichment of the economic activity mix, the improvement of spatial cohesion especially by combining important segments of the local production system with an upgraded transport and communication infrastructure, and local labour market initiatives. In other words, the strong local economic targets cannot be separated from social and political objectives and from the socio-political dynamics. The improvement of the quality of housing, neighbourhood, natural environment, associative life and political administration are as important ingredients of local development strategies as are the distinct economic targets.

Moreover, it has been stressed that local development strategies must combine economic, political and social targets, implying that development agents, cooperation patterns and instruments cannot be built upon a purely rational economic standpoint. The orthodox *agent-instrument-target* transmission mechanism does not work in most societies, and certainly not in those which have suffered from a self-reinforcing vicious circle of economic, social and political disintegration. New forms of agency must be conceived, based on different types of networks creating synergies between the public and the private, the different spatial levels of government and administration, different interest groups, as well as a wide array of professional and organizational skills which are essential to the regeneration of local social dynamics that could counter poverty trends.

All new forms of agency and cooperation could be an additional component in the making of a new type of innovation strategy rationale. Different examples from different localities that have been studied in detail by the programme clearly demonstrated that orthodox spatial development recipes inspired by the Industrial Districts or the Technology Districts models proved to be highly detached from the reality of the socio-economically disintegrated localities. Thus, diverse conceptions – of sectoral diversification, of industrial organization, of ecological sustainability – could prove to have greater relevance for innovative development in disintegrated localities than mere technology based strategies. This does not imply that technology must be excluded from local innovation strategies, but that it should be perceived as an integral part of a community based innovation strategy. As a matter of fact, this could be seen also as an early critique of the approaches such as smart growth and smart city development that emerged in the decades that followed the research.

The research project also revealed that community-based innovation strategies must rely on a strong empirical knowledge and understanding, considering the particular local characteristics and the in-built developmental potentials of the areas in question. These potentials emanate from the sectoral history of the localities, the nature of the disintegration process affecting them and the community response to this process. Of importance has also been the reading of the tendencies and the differences in the broader politico-institutional systems (national, regional, local) affecting and providing room for manoeuvre for local economic development policies in the EU Member States. Several contradictory tendencies among countries, decision-making levels and development policy agendas have been identified. Evidently strong local governance systems – capable of delivering policies supporting socially innovative development strategies – have also proved to be a critical parameter, although difficult to construct. Such difficulty has been determined by macro-economic policy conditions stipulating budgetary constraints, monetary discipline and in turn privatization doctrines. In addition, all the aforementioned in combination produced, among others, disjointed and unbalanced initiatives (by public and private agents) at the local level that further operated against the possibility of structuring a coherent and integral policy rationale. ◄

ECONOMIC STRUCTURE OF LOCALITIES UNDER STUDY AND IN RELATION TO THEIR RESTRUCTURING PROBLEMS

Types of communities under study	Peasant communities	Rural communities with light industry	Rural communities with metal and textile	1. Coal mine and metallurgy-comm. 2. Harbours	Special cases
Names of communities	Ostiglia Spatha Comarca Montes Oca Arganil	Maniago Vigevano Almeida	Urbania Roanne Mazamet Algueda	1. Rhondda Valenciennes Charleroi Dortmund S. Cardiff 2. Lavrion Elgoibar(H) Barakaldo Charleroi 3. S. Cardiff Calais Bremen Hamburg Antwerp Rostock Liverpool	Fishguard Sykies Vera Cruz Perama Girona
Specific economic problems	Infrastructure Organizational skills	Organization and management of production and markets	Idem Sectoral restructuring	Infrastructure Sectoral restructuring	

Source: Moulaert et al., 1994.

5. MIGRATION AND URBAN REGENERATION

◆

Felicitas Hillmann

The role of social innovation for urban and regional development has been aptly theorized in the innovation literature (Moulaert et al., 2005). A number of research projects have demonstrated the importance of bottom-linked collective initiatives for social cohesion. We see these groups as fundamental within the process of knowledge production. They should be approached through transdisciplinary research that reaches out to different communities of practice, interest groups, private and public actors (Miciukiewicz et al., 2012).

Even though the issue is often of utmost relevance to urban development, the social innovation project frequently missed any sharp focus on migration. This absence was addressed by the European project SOCIAL POLIS, which brought forth the work on migrant entrepreneurship and migration in relation to social innovation. This focus allowed the author to consider empirically the case of Neukölln in Berlin. At that time, the entrepreneurial activities of migrants were not seen to be of great interest for urban development. They were perceived as marginal to the rest of the society – being part of what many called "urban marginality". In contrast, our study showed that for the declining area of Neukölln migrants' activities were crucial and had introduced changes within the institutional setting. Migrant entrepreneurship and the cultural activities put forward by migrants, such as the carnival of cultures and the festival 48-hours Neukölln, were of crucial importance for the regeneration of the urban fabric. The Anglo-Saxon concept of *urban marginality*, as explanatory grid for the situation in Neukölln, was only partly right. Our research provided another twist in a context where inhabitants with migration backgrounds outnumbered those without, migration had become a central factor in urban regeneration: the number of migrant businesses was already high in that district, and the activities of migrants had already led to various changes within the local policies. In addition, migrant organizations had started to network among themselves and one little African shop turned out to become a major point of reference for all sorts of contacts. At this stage it was less the economic, and more the social innovation that contributed to a shift in overall policies in Neukölln. Migrants were in some ways responsible for large parts of the adopted bottom-up policies. Our study also revealed that it took almost ten years for these migrant-led bottom-up policies to become visible to the rest of the town.

It is still often claimed that migrants and mobility are rather marginal for our concepts of urbanity and urban regeneration. This approach might not be adequate any longer. Urbanity hardly exists without the imprint of migrant economies and activities of migrants – even if this has not been fully recognized by the mainstream society. As in other European cities, the influence and the institutional utilization of migrant activities became crucial in the 2000s, a real asset to urban regeneration and, also, commodification of the "ethnic". Today, I still make use of those ideas and work on what I call *migration-led-regeneration*, which in my view is as important as *culture-led regeneration, investment-led regeneration and community-led regeneration.* ◄

© Alessandra Manganelli

04

SOCIAL INNOVATION: PUTTING THE MARKET BACK IN ITS PLACE

6. FROM TECHNOLOGICAL TO SOCIAL INNOVATION: REHABILITATING THE STATE AND CIVIL SOCIETY IN THE DEVELOPMENT DEBATE

◆

Kevin Morgan and Flavia Martinelli

One of the most stimulating strands of territorial development theory over the past 25 years has been the confluence of innovation studies and regional development studies. As a result of this cross pollination, innovation theorists have learned to appreciate the significance of the spatial dimension, while economic geographers and regional economists have drawn upon innovation theories to help explain the processes and patterns of uneven spatial development. This confluence has generated a whole series of *place-based* innovation theories, from the *innovative milieu* proposed by Aydalot and the GREMI school in France and the retrieval of the *industrial districts* concept by Becattini and others in Italy in the 1980s (Moulaert and Scott, 1997), to the identification of *regional innovation systems* and *learning regions* (Morgan, 1997) in the 1990s, later referred to as the Territorial Innovation Systems (TIS) literature.

The TIS paradigm may have helped to establish the significance of *institutions* and the *spatial* dimension of innovative processes in development, but it did so at the cost of neglecting the *social* dimension. Another shortcoming of many TIS models was that they privileged a certain model of development (namely the STI model, based on science, technology and innovation) and extolled the role of certain actors (especially high technology firms and research-based universities). More importantly, by stressing the role of place-based assets (including social capital and *untraded interdependencies*) in explaining the innovative capabilities of certain territories, it shifted attention from the structural determinants of uneven development to local factor endowment.

In this chapter, we discuss how Frank Moulaert and other colleagues' work has played a key role in exposing the shortcomings of the conventional TIS paradigm. Furthermore, by highlighting the spatial significance of *social innovation*, Moulaert has particularly laid the basis for compelling new models of territorial development that privilege human needs rather than private profit and help to recover the agency of civil society and public sector actors in fostering more inclusive forms of development. We also point out some of the dangers involved in the recent mainstreaming of social innovation theory and use the concept of the *foundational economy* to highlight the potential of the social innovation paradigm.

TERRITORIAL INNOVATION SYSTEMS: THE MISSING SOCIAL DIMENSION

The development and conceptualization of social innovation has occurred over a number of years, particularly starting with the DG XII-funded *URSPIC - Urban Redevelopment and Social Polarization in the City* project of 1997-99, coordinated by Frank Moulaert as a critique of the Territorial Innovation Systems literature (see Moulaert, 2000; Moulaert and Sekia, 2003), and continuing with the *SINGOCOM – Social Innovation Governance and Community Building project*, funded by the European Commission within FP5 (see Moulaert et al., 2005, 2007a, 2010) and the KATARSIS and SOCIAL POLIS actions, funded within the FP6 and FP7 respectively (see Moulaert et al., 2013b).[1]

The focus of these projects was on *communities* that had been 'exposed to [...] dynamics of social exclusion [...] and on how these communities fought back in socially innovative ways' (Moulaert, 2010: 4). A theoretical framework and a transdisciplinary methodology were perfected over the years to analyze and evaluate social innovation initiatives and action throughout Europe, which integrated the debates about urban social movements and grassroots initiatives (see Martinelli, 2010 for a historical review) with studies of the

---

1. For a brief overview of these projects, see Moulaert et al. (2013a).

social economy and concerns about the dynamics of uneven territorial development that marginalized certain neighbourhoods and places.

The critique of territorial innovation systems

In a seminal article in 2003, Moulaert and Sekia presented a critical survey of the TIS literature and found it wanting because, for all the different variants, most of them framed development in highly restrictive and one-dimensional terms. Starting from this critique of TIS and their focus on technology-driven innovation and competitiveness, the article brought back on stage the missing socio-political dimension in both the analysis and the practice of local development. To overcome the reductive conception of development put forward by TIS, an alternative model – the *Integrated Area Development* (IAD) model was developed under the Poverty III programme (see Delladetsimas in section 3) which laid the foundations of subsequent conceptualization of social innovation for territorial development.

Territorial development, according to the IAD model, should be based on a multi-dimensional view of innovation, economic dynamics and community governance, and it aims to bring into focus other parts of the economy (like the public sector, social economy, cultural sector and artisan production for example), as well as community life in all its manifold forms. It is considered innovative in two ways: (1) in its more inclusive social relations of governance and (2) in its satisfaction of needs that are not satisfied by the market. In short, the IAD model helps to render visible what was rendered invisible by the narrow lens of the TIS paradigm, by rehabilitating community-based processes of social innovation and non-market agents like civic society and public sector bodies.

Social innovation and community development

The starting point of theoretical framework for place-based social innovation was the French notion of *innovations sociales* proposed in 1982 by Chambon, David and Devevey, who highlighted the role of local *collective* initiatives in fighting social exclusion and improving the quality of daily life. Other influences were the work developed by Fontan, Klein and Harrisson within the CRISES-Centre for Research on Social Innovations in Quebec (Fontan et al., 1999; Klein and Harrisson, 2006). These notions were further enriched by Moulaert in terms of recovering the insights of classical and institutional economists and incorporating the critical theoretical perspective and debates on development, the role of the state and civil society, the social economy, participatory democracy and planning.

This conceptualization of social innovation is both analytical and normative. Social innovation is defined in relation to human development and social emancipation. It is characterized by three strongly interconnected features (González et al., 2010): (1) the satisfaction of human needs that are not answered in the current state of affairs, whether they are material or immaterial needs; (2) the capacity to change established social and power relations among the civil society, the state, and market actors, both within the community and between the community and the outside world; and (3) the empowerment of excluded groups, via the collective (re)construction of identities, capabilities, cultural emancipation. Social innovation is, therefore, conceived both as an *outcome*, namely the actual satisfaction of needs, and as a *process*, whereby excluded groups are empowered, knowledge is created, and local capabilities are enhanced.

The application of the social innovation approach by Moulaert et al. (2010) and colleagues in the various projects deployed over the years involves a *postdisciplinary* stance, meaning it requires a *multidisciplinary* analytical approach to valorize and integrate the diversity of disciplinary insights (management sciences, economics, sociology, anthropology, political sciences, public administration), but also a *transdisciplinary* type of research, that is one in which academic researchers interact with practitioners and local actors. 'Reflexive' transdisciplinarity (see Jessop et al., 2013) especially informed the action-oriented research methodology applied in the KATARSIS and Social Polis projects.

The *spatial* dimension is a key aspect of Moulaert's approach. Territory is both an analytical tool and a field of action (Van Dyck and Van den Broeck, 2013: 133). The community is not defined by ethnic, class or other socio-cultural parameters, but as a *spatialised* set of often very diverse groups and actors, who interact in a given space and deploy their action. It is, indeed, at the *local* scale that community dynamics are best revealed and action can best deal with the diversity of agents and agencies. The main thesis is that the local scale is the 'pivotal site for initiating and implementing social change that may ripple through the city' (Moulaert, 2010: 5). With globalization, in fact, the national space has lost prominence as the privileged scale of state intervention. And with the affirmation of the neoliberal paradigm in the 1990s, many domains of public life have been commodified and citizens' rights reduced to market transactions, contributing to the privatization – and de-politicization – of the governance of public spaces and activities.

But although the local is the main focus, Moulaert et al. (2010) are well aware of interscalar relations and the need to take into account the broader macro-scales. They stress at the outset that 'the word "local" refers to an articulated spatiality, in which the local is the site of existence of a proactive community, but also a node in a complex geography [...] transcending the confines of the local' (Moulaert, 2010: 11). They also warn about the dangers of placing an 'exaggerated faith in the power of local level agency and institutions to improve the world, thereby ignoring or disavowing the inter-scalar spatiality of development mechanisms and strategies' (González et al., 2010: 50), and they are well aware that higher-level state and corporate actors tend to dump on the local scale the costs of globalization and austerity.

Innovation at the local level is thus an "empirical entry point", a "window" into a broader multi-scalar picture. According to Moulaert et al. (2010), working at the community scale provides the nexus between, on the one hand, the redefinition of everyday social life in the community and, on the other hand, the broader struggle for democracy and rights. There is a tight dialectic between community- and society- building, since solidarity-based community-building is not possible without guaranteeing generalized citizens' rights and without the transformation of governance institutions and practices through a redefinition of state-civil society relations.

Beyond localism: the limits and potential of social innovation

The spatialized conceptualization of social innovation in the context of community development has rehabilitated the socio-political dimension in the debate about local development and territorial innovation systems. In this approach, territorial innovation and development are not just about technological-economic processes, but also about social and human needs; not just about competitiveness, but also about solidarity; not just about state and corporate players, but also about civil society action and participatory governance. These features continue to set apart the work by Moulaert et al. (2010) from the many mainstreamed versions of social innovation that have developed in the last decade or so and have been sponsored in many policy circles (see for example Young Foundation, 2010; OECD, 2010; Mulgan, et al., 2007; BEPA, 2011).[2]

And yet, Moulaert et al.'s approach still exhibits a number of weaknesses. Despite the repeated calls to a multi-scalar contextualization, it remains somewhat trapped in a *localist* perspective. While it acknowledges the influence of broader structural forces in causing social and territorial exclusion at the local scale, the transformative potential of socially innovative actions at the community level encounters difficulties in working up-stream and changing social and power relations at higher societal and government scales.

In fact, the social innovation approach does not really challenge the *structural* trends that determine social and territorial exclusion. Reference to the *social economy* as an alternative way of organizing things remains subordinate to the capitalist market economy.

---

2. In these mainstream versions, which rely on a mostly managerial and economistic approach, social innovation is seen as a universal — de-territorialized — recipe to solve the problems generated in many places by the financial crisis, be it unemployment, marginalization, urban decay, or cuts in welfare services. The social economy itself is reductively conceived in micro-economic organizational and efficiency terms. In this business-oriented policy discourse, which promotes the privatization of public services and the substitution of collective logics with competitive ones, social innovation is emptied of its emancipatory dimension and reduced to a means for allegedly increasing the efficiency of the public sector and in fact creating greater opportunities for profit-making services (for critiques, see Martinelli, 2012; Jessop et al., 2013).

The limits of social innovation at the community scale

While it is the most appropriate level for the revelation of needs and the experimentation of new ways to respond to those needs, the local scale also represents an inherent limit to the diffusion and up-scaling of social innovation beyond the community. This difficulty was already acknowledged in empirical accounts of social innovation (Martinelli et al., 2010). Although emulation and replication of socially innovative local initiatives through regional, national, or even international networking is one way to bring about the spatial diffusion of social innovation, the actual institutionalization of successful initiatives at higher-than-local scales has occurred only sparingly, putting pressures on the long-term sustainability of many initiatives.

Another danger of the social innovation approach is – as already mentioned – its instrumentalization in mainstream discourses, whereby the social economy and bottom-up initiatives must take care of needs that were once addressed by public welfare provisions. Former universal social rights are now retreating, and services must be co-produced or even produced by third sector organizations and the social economy, with shrinking public support (Martinelli, 2012; Martinelli et al., 2017). Therefore, not only are they supposed to substitute for previous entitlements, but they are also highly vulnerable to the creed of austerity, a creed that aims to cut the deficit and shrink the state.

Another limit related to the community or local scale is that in many places, such as extremely marginalized communities, mobilization simply does not occur because of lack of resources, be it knowledge, social capital, or leadership. This problem exacerbates territorial disparities and also justifies top-down assessments that tend to "blame the victims", because they are unable to take their destiny in hand.

For all these reasons, even if it brings in the missing social dimension, social innovation at the local scale does not escape the main criticism of the TIS paradigm: the fate of places and their capabilities to engineer change depends on their local resource endowments. Even if new values, dimensions and relations are mobilized to ensure an inclusive development model, places remain marginal and excluded when the necessary local resources are wanting. This problem stems from not taking into account the broader structural forces that drive development and – in the present absence of redistributive public policies, captive to neoliberal recipes – intensifies spatial inequalities. In fact, it indirectly endorses the neoliberal paradigm of endogenous development based on competitive factors and its policy corollary of providing public resources through competitive tendering, which end up pitting places against each other.

Finally, social innovation at the local level, however networked and institutionalized, does not really address contemporary social and power relations, characterized by the dominance of financial capital, real estate power brokers, and predatory transnational corporations, which have stymied the competence and confidence of national governments. Civil society and social economy actors can organize alternative local production and consumption models, but these models have little chance to modify dominant power alliances and exploitative relations. This can only be achieved if changes are implemented at the higher – that is national or international – governmental scales via regulatory enforcement.

From social innovation to system innovation: the foundational economy

To overcome these limits, locally-focused social innovation strategies need to be calibrated with and bolstered by *national level* support systems that can help to sustain social innovation movements. What is needed above all else is a link between the local and the national or supra-national level – a mechanism that connects socially innovative practices at the local scale with national (and EU) public policies, reinforcing each other. This link is not easy to create, since public policy by the national state has been stymied by the *age of austerity* and EU Cohesion policy is in thrall to the neoliberal mantra of competition, entrusting localities with the responsibility of formulating development strategies and bidding for resources.

But the scope for social innovation can be broader than its elective local scale, especially when it is calibrated with the *foundational economy* (Bentham et al., 2013; Morgan, 2015). The foundational economy (FE) can help to transform social innovation from a concept tethered to social enterprise at the margins of society and economy to a set of activities that are central to the working of national society and economy - activities that collectively account for up to 40% of employment and value added in most countries.

The FE includes sectors such as social services (education, health, care), utilities, transport and communication, retail (especially food, but also gasoline and banking), food processing, that are distributed throughout national spaces following the population they serve and operate in markets and sectors that are relatively sheltered from global competition because of state regulation and/or outsourcing. Some of these sectors are also typical domains of socially innovative initiatives (for example social services, food production and distribution), but they remain localised experiments. The FE conceptualization provides the link between local organizing and national policy.

The driving forces behind the FE are a combination of municipal activism (as sub-national governments begin to counter the neoliberal policies of privatization and outsourcing) and civic engagement (as civil society organizations seek to become co-producers of local policies), but geared towards national regulation and policy, thereby linking local action with a national political strategy. The FE can provide a strong leverage to the state, precisely because it operates in sectors where the state is already present, but where it has waived its power (such as its power of regulation) or has been unable or unwilling to deploy it (such as its power as purchaser) to meet human needs.

CONCLUDING REMARKS

Social innovation has brought innovation theory to a new plane and, most importantly, has enriched spatial development studies with a hitherto missing dimension - that of the *socio-political*, involving initiatives for the acknowledgment of human needs, the empowerment of excluded groups, and changes in established relations among actors, especially at the community level.

The social innovation paradigm highlights the significance of other models of development (based on the social economy and the public sector), radically different metrics of development (based on social need not private profit), and new categories of agent (municipalities, public sector bodies and civil society organizations). These new models/agents are becoming increasingly relevant in an age of societal challenges (climate change, renewable energy, sustainable mobility, dignified eldercare, food security), all sectors in which the state looms large and in which citizen practices as well as consumer behaviour can play an important role for societal change.

The community level – be it a neighbourhood, a municipality or a rural area – is a key scale for the social mobilization of citizens, economic actors and institutions and for the implementation of socially innovative practices. Far from being confined to the local level, however, social innovation can be translated into *system innovation* if the mundane activities of the Foundational Economy – which are present in every community – are mobilized to satisfy human needs that are increasingly under threat from a pre-Keynesian ideology of austerity.

To summarise: The Foundational Economy is a compelling vehicle for mainstreaming social innovation, thereby avoiding the localist trap in which social innovation activity is often caught. The great merits of the Foundational Economy as an alternative socio-economic strategy are threefold:

1. it embraces the *national economy* and therefore it has traction at the macro/structural level;

2. it reinstates both the *civil society* and the state as actors in the development debate because citizen engagement as well as political mobilization are equally necessary to meet societal challenges in a democratic and sustainable fashion; and

3. its primary focus is on the satisfaction of *human needs* rather than the extraction of private profit.◄

7. POLITICAL ECONOMY AND REGULATION THEORY

Stijn Oosterlynck

Political economy is a key perspective around which the critical turn in social innovation for political transformation revolves, especially in terms of regional and urban development (Moulaert, 2000; Swyngedouw et al., 2002). Political economy highlights how processes of socio-economic development and state actions and restructuring are mutually implicated. When Fordism entered a structural crisis in the 1970s, many critical scholars turned towards political economic approaches to analyze the deep transformations of capitalist societies. One popular approach at the time was that of the Regulation Theory. It combines a Marxist political economy perspective with institutionalism. The former sees capitalist development processes as crisis-ridden and hence unstable, while the latter focuses on the institutions that generate social order and stability.

Regulation Approach seeks to understand why capitalism does not succumb to its own internal contradictions and survives crises. It rejects neoclassical ideals about human beings behaving like rational and self-interested individuals and capitalist markets tending towards equilibrium. Regulationists are concerned with reproduction: how are capitalist social systems reproduced, given the crisis-ridden nature of capitalism? The crisis of Fordism in the 1970s made abundantly clear that reproduction is not automatic and led regulationists to analyze reproduction in terms of regulation. Regulation refers to the social norms and expectations, the formal and informal institutions that (try to) make subjects act in accordance with the reproduction requirements of capitalism. Some of the seminal works on the geography of flexible production systems (see Moulaert and Swyngedouw, 1989; Moulaert and Swyngedouw, 1991) have inspired a range of mainly UK scholars in further exploring the spatial dimensions of emerging post-Fordist modes of social regulation (Jones, 1997; Tickell and Peck, 1992).

From the 1990s onwards, the emerging focus on integrated area development led to the search for alternative modes of socio-economic development (Moulaert, 2000). The concept of social innovation allowed critical researchers to move beyond the concerns about flexible production, market-driven innovation and high-tech entrepreneurship in the post-Fordism literature, to explore more progressive socio-economic development trajectories. From that vantage point, social innovation is very critical of regulationist inspired scholarship for presenting neoliberalism as invincible (for example Moulaert et al., 2017) and not leaving much hope for alternative models of political-economic development.◄

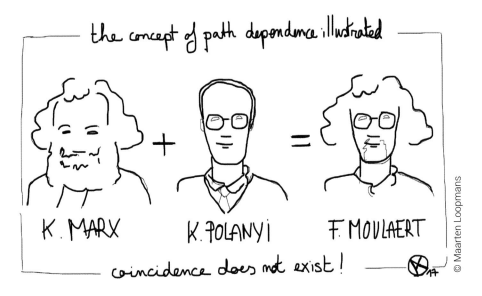

the concept of path dependence illustrated

K. MARX + K. POLANYI = F. MOULAERT

coincidence does not exist!

8. SOCIAL INNOVATION: A POLANYIAN REINTERPRETATION

◆

Maarten Loopmans and Chris Kesteloot

In the literature, social innovation is often presented with clear political, as well as analytical goals and it is as much a prescriptive as a descriptive concept. Its general definition has three components: (1) satisfaction of basic needs through (2) the reconfiguration of social relations and (3) political empowerment (Moulaert et al., 2013b). Despite this analytical clarity, it leaves us in a confused (and often confusing) conceptual state (Moulaert et al., 2013a:14). Which and whose basic needs are to be addressed? Which social relations are supposed to change? In what direction? What does political empowerment entail for whom? These questions remain unanswered in the basic definition of social innovation.

In this short piece, we argue that the writings of Karl Polanyi allow us to develop the critical and prescriptive dimension of social innovation. The basic concepts Polanyi deploys to understand and evaluate the *Great Transformation* that capitalism brought about in the long nineteenth century, being the *modes of economic integration* and *fictitious commodities*, could serve us to provide normative directions for social innovations (Polanyi, 1944).

MODES OF ECONOMIC INTEGRATION

The basic tenet of Polanyi's discourse is that all historical societies have used a combination of different systems to regulate production and distribution of humanity's means of existence. The combinations are manifold but use only three basic types of exchange. Markets based exchange on prices determined in an impersonal way by confronting demand and supply. Reciprocity is a long-term exchange relation between people (and sometimes social groups), commanded by needs and the capacity of others to meet these needs. Redistribution involves a central institution that collects goods and services to redistribute them.

Polanyi reveals how each mode of integration requires its own social institutions and rules of behaviour. Market exchange implies atomistic individuals who pursue their own interests. These individuals must be autonomous in their production and consumption decisions, to be able to respond to the signals of the prices. The "invisible hand" of the market brings all actors in competition with each other, rather than bonding them within institutions. Redistribution implies centrality and hierarchy, providing a central institution with the political authority to decide about the collection and redistribution of goods in the name of all participants. In contrast, in reciprocal exchanges, individuals take decisions balancing needs and possibilities within a social network. This network connects people among which enough mutual knowledge and trust can be built up to allow for these exchanges. Because a long-term balance has to be achieved between what someone gives and receives to and from others in the network, social relations are symmetrical and egalitarian. Typically, reciprocity occurs within networks of friends, kind or tightly knit communities. In his historical and anthropological work, Polanyi describes how societies differ in the mix of modes of economic integration they deploy, implying that social and economic history can be described in terms of changes in this mix. The *Great Transformation* of capitalism, he implies, consists of markets acquiring an ever-greater share of control over the production and distribution of the means of existence, to the detriment of redistribution and reciprocity. But *market dominance* is about more than expanding the field of operation of market exchange. Polanyi describes how as a dominant mode of economic integration it also imposes its logic to the institutions underpinning other modes of economic integration (Jessop, 2001: 219).

FICTITIOUS COMMODITIES

The most dramatic and contentious expansion of market exchange occurs with what Polanyi calls "fictitious commodities", namely labour, land and money. None of them is produced for sale and profit. Nevertheless, they can be *commodified*, that is privately owned and exchanged through markets. In his own words: '... labour and land are no other than the human beings themselves of which every society consists and the natural surroundings in which it exists. To include them in the market mechanism means to subordinate the substance of society itself to the laws of the market' (2001: 75). The condition of this commodification is the destruction of the social and political institutions that regulate the use of labour, land and money. But it also deeply affects the fate of men and nature.

Labour (capacity) is contained in human bodies and producing and maintaining labour capacity entails producing and maintaining human life. Its price, resulting from supply and demand, enables their producers to purchase their means of existence as commodities on the markets. But that price, the wage, can plunge under the level needed for survival. Hence commodified labour affects individuals' right to live and imposes a market logic on the social institutions necessary for its reproduction. *Land* is not produced and the use value of a plot of land is inherently a social product. This renders the commodification of land into a vehicle for unproductive and socially disruptive speculation and exclusion. The search for land uses that yield the highest price, the land rent leads to the very destruction of nature. As land (and the seas) is also the container of nature and all resources needed to sustain human life, the alienation and exclusion of people from access to land equals their annihilation. Similarly, money as an instrument of exchange obtains its use value from the social institutions producing it. If money is transformed into a commodity with a price, the interest rate, based on supply and demand, it too becomes an object of unproductive speculation on future value, disrupting its contemporary role in exchange. Just as for land, the consequences of such speculation are inflation and crises with rapid devaluation of money and the dramatic loss of purchasing power for firms and people who use it as a mere means of exchange and savings.

DISEMBEDDING TODAY

Central to Polanyi's thesis is the socially disruptive impact of the disembedding of exchange from social institutions, and the consequential fictitious commodification of land, labour and money. A complete market economy is 'a stark utopia which could not exist for any extended period without annihilating the human and natural substance of society' (Polanyi, 2001: 3). Hence, individuals and groups experiencing the disruptive character of fictitious commodification in their everyday life necessarily react. Polanyi showed interest in these various acts of resistance and survival from those affected by the destructive forces of the market; but his emphasis is on how these social innovations add up to a broader *act of self-defence* from society to re-embed the economy and shift the balance between market exchange, reciprocity and redistribution.

Polanyi seemed to consider the development of the welfare state as the *end-state* of this double movement of disembedding and re-embedding. Unfortunately, however, the struggle continued, and it seems that the pendulum has swung back towards market dominance since the late twentieth century crisis called neoliberalism. To re-establish profit margins and opportunities for capital accumulation, capitalists have revamped disembedding and commodification in the past half a century. A myriad of strategies has been developed to deepen the exploitation of labour and extend the fields of operation of the market to arenas hitherto regulated by social institutions. The figure below attempts to summarise them.

The extension of the fields of operation of the market system has been spurred by globalization (expansion into exchanges, lands and populations outside capitalism through revamped imperialism and the disintegration of the communist bloc), financialisation (commodification of money), flexibilisation and restructuring of the state (including the imposition of a market logic on redistributive functions). Strong arguments have been produced to include knowledge as a fourth fictitious commodity and indeed the commodification and financialisation of knowledge is on its way, unleashing similar destructive forces on knowledge production (Jessop, 2007).

The second response, deepening the exploitation of the workers, has been enabled by removing political and social controls over the access, use and remuneration of labour force, that is the recommodification of labour. Geographical competition, relocation of economic activities, outsourcing and immigration have greatly expanded labour markets and loosened the grip of societal and communal institutions on the reproduction of labour.

Fiscal and social security reforms resulted in a significant reduction of the share of indirect collective wages (redistribution) in the reproduction of labour and increased dependence upon supply and demand based labour markets to sustain human life. In combination with job insecurity, unemployment and the deregulation and informalisation of job markets, this has greatly increased social needs in terms of multifarious inequalities, polarization and impoverishment.

◆

RESTORING THE RATE OF PROFIT IN THE NEOLIBERAL ERA: DISEMBEDDING AND RECOMMODIFICATION OF LABOUR

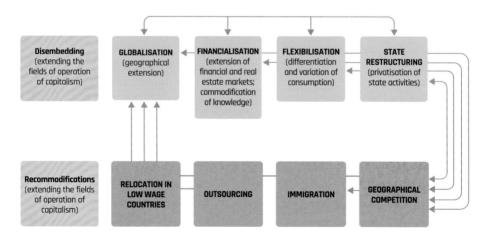

Source: adapted from Kesteloot (2013:11)

SOCIAL INNOVATION AS RE-EMBEDDING

Like social innovation theorists, Polanyi does not wait for grand revolutions to change the world but is interested in small incremental changes. But he did indicate a clear direction for these changes: away from market dominance, towards greater social control over economic transactions. It made him, in the 1930s and 1940s, a champion of democratic socialism and a source of inspiration for the welfare state. Could his analysis still provide guidance for socially innovative strategies under and against neoliberalism today? And who could be the initiators of such progressive social innovation leading to socialist democracy in which markets are again subdued to social goals and democratic control?

Looking at social innovation through a Polanyian lens, one can be more specific about the three dimensions of social innovation (needs fulfilment, changing social relations and empowerment), as identified by Moulaert et al. (2013b). The kind of social innovations Polanyians would look at are those that react to the *needs* created by neoliberal commodification. Reactions coming from those affected by and trying to redress the effects of unemployment, informalisation of labour, job insecurity, the decline of collective wages, but also the destruction of nature are social innovations that have the potential of pursuing effective and progressive social change. Secondly, social relations should be changed to provide alternatives to the logic of the market (see also Moulaert and Nussbaumer, 2005). Such alternatives can be sought in an expansion and renewal of the institutions and *social relations* underpinning redistribution and reciprocity, to allow for a greater share of exchanges regulated through these two spheres. Additionally, a renewal of social relations can also curb the influence of the market logic in redistributive and reciprocal institutions. Crucially, these innovations should address the sharing and the governing of labour, land, money and knowledge. Thirdly, *empowerment* means *democratisation* of the economy. Progressive change towards an embedded economy is to be expected from those feeling the needs, not from economic elites; although Polanyi seems to provide for a role of political elites in such social innovations, in collaboration with grassroots social movements in the scaling up of micro social innovations to the societal level (Dale, 2016).

If social innovations are to constitute a re-embedding movement, they should provide alternatives to today's neoliberal system. They should swing the pendulum again, away from the prominence of market exchange and the predominance of the market logic. They should challenge this logic and provide for needs on a different basis. Examples do exist. Many of them develop in the margins of the systems, in spheres where the pressure of the market logic has penetrated to a lesser extent. Thinking out-of-the-box and thinking forward about social innovation means looking for what happens in the periphery of the capitalist system and working together to accumulate and congregate these initiatives into a broader societal transformation towards democratic socialism.◄

05

Gare Lille Europe © Constanza Parra

THE GLOBALIZED CITY

9. PRODUCER SERVICES AND REGIONAL DEVELOPMENT IN THE AEGEAN WITH PARTICULAR FOCUS ON LESVOS AND CHIOS

◆

Pavlos-Marinos Delladetsimas

This research programme was developed with a focus on two major islands of the North-eastern Aegean Sea: Lesvos and Chios. The island of Lesvos has a tradition in agriculture especially in olive-oil production. Historically it has also developed a strong industrial base, which has been dramatically curtailed in the post-war period. The production structure of the island of Chios is characterized as "the mastic network": this is a unique produce of the island, based on a resinous substance from the mastic tree which grows only in the southern part of the island. The product is used in the pharmaceutical and perfume industry, varnishes and paints, and in food and drinks, among others.

The research constituted a continuation of work on local development analysis, making a plea in favour of a historical and institutional approach (Moulaert, 1996; Moulaert and Delladetsima, 1998). Without such an approach it is not possible to root new development ideas and practices in the territorial potential of localities. It thus shows how essential it is to get a grasp of the history of the socio-economic structure, the human, natural and technical development resources, and the institutions that in the past favoured or checked development. Institutional *path dependency* offers the missing link in the study of regions and localities with a *strong* history (Lambooy, 1988; Moulaert and Delvainquiere, 1994).

This research integrated theoretical elements of historical analysis, local development theory and insular economics into a sound theoretical framework of local development in island communities. The theoretical synthesis also took into account the elements of economic anthropology, interrelations between economy (agriculture, artisan production), polity and trade, the role of the hinterlands, the role of the transport sector in the spatial definition of markets, and infrastructure for logistics. The work systematically identified links between traditional production systems (olive oil in Lesvos and mastic in Chios) and service production networks taking different network forms. It revealed the slow but significant opening of these traditional sectors for contemporary producer services, embracing local suppliers and external agents especially for consulting training. At the conceptual level the work clearly demonstrated that local development, as driven by more routine or less advanced producer services, differs from the model put forward in the high technology and the advanced producer services *learning region or locality* (Morgan, 1997). At the empirical level it introduced an innovative type of empirical research – adjusted to insular economies – that had not before been realized in the Aegean economy or internationally. It thus revealed new models of *intermediate innovation*. These models emanated from a distinct local character and proved to be highly effective within a context of insularity, isolation and the adverse developmental trajectories of the two islands.◄

Location map of Lesvos and Chios Islands (Source: Harokopio University)

Production phases in olive oil production.

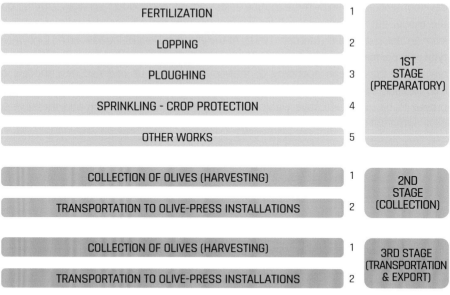

Source: G. Gavrilakis - G. Georgiadis (1996), Thesis; University of the Aegean. Dpt. of Environmental Studies

The production phases of the mastic production process.

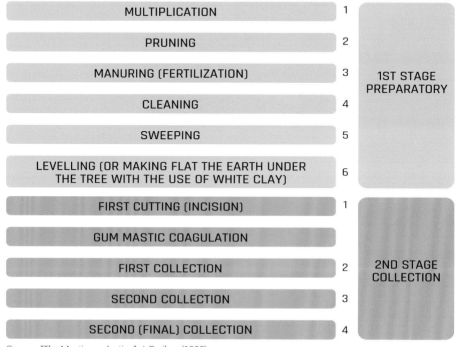

Source : "The Mastic production", J. Perikos (1993)

10. NAPLES

◆

Jonathan Pratschke, Lucia Cavola and Enrica Morlicchio

Naples – the industrial capital of Southern Italy until the end of the 1980s – is attributable, at least in part, to the way in which the city condenses aspects of Southern European society. At the same time, its specificities – its rich history, creativity and cultural vibrancy, but also its widespread poverty, political clientelism and organized crime – make it a rather extreme case, and one which is particularly relevant to the study of social polarization. During the 1990s, as Moulaert and colleagues were theorizing the links between social needs and social innovations and the challenges that the latter pose for governance, the Neapolitan case provided an opportunity to assess the resulting ideas in a context characterized by such contrasting trends.

These themes and issues were explored in the course of a series of European projects, including SINGOCOM and URSPIC. SINGOCOM included a detailed analysis of the

Quartieri Spagnoli and the Piazziamoci network in Scampia (Moulaert, 2007), whilst URSPIC studied the development of a new business district in an ex-industrial area (Rodriguez and Martinez, 2003). These projects shed considerable light on the socio-spatial dynamics that characterized social and economic development in the centre, in deprived peripheral housing estates and ex-industrial sites.

This comparative research generated a number of interesting findings. It showed that the pronounced forms of social and spatial polarization observed in Naples are driven by rather different mechanisms to those found in Northern Europe. Secondly, it revealed that large-scale urban development projects – where these occur – are driven less by neoliberal policy prescriptions than by "traditional" corporatist interests. Thirdly, socially innovative projects in Naples tend to take shape largely outside and in opposition to the local state. Fourthly, there is a relatively extensive network of voluntary bodies which play an important role in tackling social exclusion at neighbourhood level. Finally, the local state is embedded in particularistic forms of interest representation which generally preclude popular participation.

These results focus attention on the role of power and governance systems in facilitating or obstructing innovative ways of satisfying social needs. A 2007 article draws together these strands with an aim to theorize the mechanisms underlying social exclusion in European cities using a unitary framework (Moulaert et al., 2007c). It shows that the same factors are crucial in all cities, although they can function in different ways and affect different groups. In Naples, reliance on informal family and friendship networks, informal or illegal (but not necessarily criminal) activities and neighbourhood life generate quite a conservative model of self-reliance, characterized by lack of trust in external institutions. Pratschke and Morlicchio (2012) make similar arguments in their analysis of social polarization in European cities, carried out as part of the SOCIAL POLIS (http://socialpolis.eu) project.

The end result is a cyclical pattern in which bottom-up social innovation is widespread in Naples, but largely confined to specific contexts due to the shortage of resources and the unresponsiveness of the political and administrative systems. Rather than feeding into a virtuous cycle of reform and social learning, innovative actions often become part of a vicious cycle of social isolation and adaptation. The weakness of demand for labour across the city and its segregated and polarized nature make it difficult for individuals from lower-class backgrounds to get ahead by studying, and it is equally difficult to achieve results by protesting. Spontaneous political mobilizations – such as the long-lived *organized unemployed* movement – tend to degenerate, becoming dependent upon political patronage and ad hoc forms of assistance.

Although Moulaert et al. (2007c) do not reach firm conclusions regarding potential solutions, they make several useful observations. Firstly, they emphasize the importance of spatial scale and the need to integrate different levels of analysis. This appeal to integrated and scale-sensitive approaches is associated with an allegiance to macro-level political economy perspectives as well as micro-level studies of innovation and participation. In the Neapolitan context, this suggests that localised forms of social innovation cannot lead to sustainable ways of satisfying basic needs unless they are accompanied by innovative policies, mobilization and collective participation at higher scales. This implies that the mutually reinforcing structures which reproduce poverty and social exclusion in European cities require an integrated and *global* response. ◄

11. EURALILLE AT THE SERVICE OF AN UNACCOMPLISHED LILLE METROPOLITAN AREA

Thomas Werquin

The Euralille, a large scale urban development project, started in the early 1990s under the aspirations of Pierre Mauroy, the visionary mayor of Lille and former prime minister of France. Given the central location of Lille between three capital cities of Europe (Brussels, London and Paris), Mauroy saw an opportunity when it came to connecting these cities via high-speed railway links. To him, Euralille could play the role of a *tertiary turbine* and attract corporate headquarters to help with the economic transition of the region from the 1960s' post-industrial decline.

The conception of Euralille happened during difficult times with the international real estate crisis besides the overspending related to the creation of new high speed train station (Lille Europe) in the centre of the city (Moulaert et al., 2001a). Notwithstanding initial difficulties, Euralille became a highlight of the Lille metropolitan region's revival with its highrise office buildings, numerous services, apartment blocks with 5000 new residents, and as many jobs. In 2016, various real estate projects were still under construction while more were planned, including a large tower building of 40000 m² office space, adding to the stock constructed in the past twenty-five years.

Together with the regeneration of the old city centre, the first driverless metropolitan railway system in the world and the city's large museums, Euralille has undeniably contributed to improving Lille's image, and its transformation from an industrial city in decline to a modern touristic and tertiary metropolis. Contrary to its otherwise modest and fatalist nature, Lille showed a high ambition when applying in vain for the organization of the Olympic games, followed by a successful campaign to become European Capital of Culture in 2004.

AFTER EURALILLE...

Within twenty years of transforming the face of Lille, Euralille inspired other grand projects such as Euratechnologie, dedicated to digital technologies, which emerged in the heart of an immensely regenerated former textile region. In Roubaix and Tourcoing, two major towns of the Lille metropolitan area that also suffered from the textile industry's decline, considerable investments contributed to the gentrification of the city centres, including the construction of an aquatic complex, a museum and other public services oriented towards artistic creation (La Condition Publique) and innovation (Centre Européen des textiles Innovants, l'Imaginarium, and so on). Beyond these public investments, the Lille metropolitan area has also seen many successful family-based, multinational and blue-chip companies, including Auchan, Décathlon, Leroy-Merlin, Bonduelle, Promod, Roquette, and so on.

Still, from some basic data, including the evolution of private employment and demographic growth, a less favourable dynamic can be observed in the Lille metropolitan area, as compared to other French metropolitan regions such as Lyon, Marseille, Bordeaux and Toulouse.

A FAVOURABLE BUT INSUFFICIENT DYNAMIC

The Lille metropolitan area and its surrounding region have encountered an economic catastrophe, the effects of which cannot be easily remedied. As such, positive signals hardly compensate the huge loss of employment in the industrial sector.

Public investments meant to stimulate service sector employment, have produced some positive impacts, among others in Euralille and Euratechnologie. However, compared to the ratio of unemployment in the Lille metropolitan area, the 3000 regained jobs in Euratechnologie are in the end insufficient. It seems that the strategy of government-supported large scale urban development projects certainly produced very

EVOLUTION OF SALARIED PRIVATE EMPLOYMENT IN LARGE FRENCH AGGLOMERATIONS

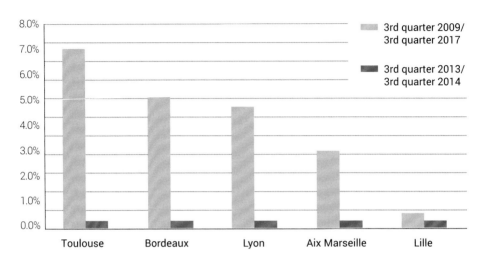

Source: ACOSS / Graphique: AxeCulture

positive effects, but these were also very costly and had comparatively limited socio-economic impact on the region.

One of the explanations of the difficulties in the Lille metropolitan area is without doubt situated in the complex territorial governance. This includes the ways in which multiple administrative layers get caught up, as well as the competition between the numerous prerogatives which the hundreds of municipalities try to maintain, in various public policy fields (land, housing, economy, and so on). This hinders the elaboration and implementation of a coherent strategy for a territory encompassing two million inhabitants and the thirty kilometre long old mining area. Furthermore, the cooperation of metropolitan authorities with the neighbouring regions is either weak or non-existent. Consequently, the new Louvre museum established in Lens, some forty kilometres from Lille, does not directly benefit the latter.

The territorial governance of the Lille metropolitan area thus seems to constitute its biggest handicap. The city of Lille, with its 230 000 inhabitants, does not have sufficient critical mass to play a real leadership role in a multipolar territory in need of nothing less than a revolution to leave behind its industrial past and fully exist on the national as well as the European scene. ◄

12. A POSTCARD FROM LONDON

Michael Edwards

L
ondon is a big city by European standards but that is as precise as one can be. Its administrative area was already too small when defined in 1965 and becomes ever more so as its growth sucks commuters from much of England and migrants from everywhere. We know the importance of multiscalar relationships, though, and live with very distinct and localised economic, social and political experiences in the cities, towns and villages which make up our country: various capitalisms surviving under one Queen (the rentier par excellence) but a country increasingly financialised and divided.[1]

---

1. A longer text with more emphasis on housing, and with references and links, is a free download: Edwards, Michael (2016 April) The Housing Crisis and London, in Special Feature on London edited by Anna Minton and Paul Watt, City, 20(2): 222-237, open access at http://www.tandfonline.com/doi/full/10.1080/13604813.2016.1145947

The dominant discourse is so familiar. City leaders, (almost all of) the political parties, policy communities, professions and mainstream media are proud of its rate of population and GDP growth, its prowess in fostering innovation and entrepreneurship, the "light touch regulation" of its financial, housing and labour markets, its cultural richness, its universities and its youth. A wonderful place; the engine of the nation.

Policy is crafted to sustain and extend this pre-eminence, with much reliance on the benefits of agglomeration as a convenient and reassuring rationalization. The co-location of state and diplomatic functions, finance, business headquarters, elite universities (including Oxford and Cambridge an hour away) and leading cultural institutions has been a winning combination. The magic might even, we are told, carry us through our separation from the European Union. It can be presented also as green: public transport is elaborate, expanding and popular, we have reversed the growth of car use and cycling is booming.

But London is a poverty machine as well as a wealth machine and has been for centuries, harvesting the value produced under slavery in the former empire and in the post-imperial world, exploiting its resident working class in making the coffee, cleaning up, doing the caring, building, driving and security to support the growth. So, it is a city of growing inequalities and it's not so green either: its road space is increasingly filled with diesel vehicles delivering online orders and ferrying passengers who summon them by apps. The air is illegally toxic, and we don't even count our massive use of air travel and container shipping in computing our pollution load.

Much of this could be said of other cities but there are some distinctive London (or British) features in our local experience.

Above all Britain embodies the strong survival and periodic renewal of the privileges attaching to land ownership. Monarchy and aristocracy were never abolished, and the early evolution of capitalism benefitted from the channelling of old landed wealth into capitalist enterprise – in the expansion of a slave-based empire, in the innovations of factory and mining production and in speculative urban development and infrastructure. Land owning interests have retained, through all this, powerful privileges in taxation, their contractual relations with tenants, inheritance and political representation. The privatization of common land in earlier centuries has a natural continuity with today's enclosures of public space, commodification of collective assets and subordination of public planning to private profit. All of this has generated great contradictions along the way as private land ownership has blocked and distorted the evolution of infrastructure and cities, prevented the efficient housing of the population and starved local administrations of revenues.

Modern London is substantially a product of successive waves of speculative investment, but also contains the products of important class struggles in the form of extensive social housing, mainly distributed through the inner neighbourhoods where left-wing

Photo: London City Hall, Gypsies and Travellers challenge Mayor Boris Johnson's draft London Plan © LGTU

local authorities built workers' housing in the twentieth century. This has given inner parts of London a rather fine-grain mixing of social class and some inoculation against rapid transformation: a distinctive feature of the city and one which we had rather taken for granted.

The other important and distinctive inheritance is the planning system established after World War II as part of the social democratic settlement and the policies and practices which developed it in the subsequent decades. In particular, London is surrounded by a green belt, now merging into other restrictive designations of open land which extend far into the surrounding regions, preventing lateral urban growth. And within the urban areas we have many restrictive designations protecting neighbourhood character, architectural interest, views and landscapes. The market in housing is also a market in proximity to these amenities, to the best schools (in a highly unequal system) and environments. A few of us argue about the relative importance of monopoly, absolute and differential rent but we all agree that rent is a massive allocator of the social product.

It is in these specific London conditions that housing market demand has surged. It has been a combination of population growth, income growth for the rich who then tend to acquire more housing, subsidy and policy support by governments for expanding ownership and capital accumulation — and all that backed by three decades of credit expansion. Overall this has been a financialised boom in house prices. Affordability falls and the proportion of households in owner-occupation, which had risen since 1918 and peaked in the 1990s, has fallen as more dwellings are switched to private renting — a tenure form almost completely unregulated and highly insecure for tenants. More and more households are driven to rent privately at almost all income levels: better paid workers who cannot afford to buy and poorer workers who would, in former times, have entered social housing. The social sector has shrunk steadily through privatization and is now rapidly eroding as many housing providers raise their rents closer to market levels. In real terms London earnings are among the lowest in the UK and have recovered more slowly than other regions since the credit crunch of 2007/8. London workers are thus simply unable to compete in this bloated market. That contradiction had been bridged by Housing Benefit, a part of the social security regime which government capped in a desperate attempt to contain its escalating cost. Wages remain low and static for much of the population while rents continue to escalate. The outcomes are an accelerated displacement of people to cheaper areas — often far from London — growing overcrowding, broken and dispersed communities, ill health and disruption of schooling. Mainstream economists point out how well our unregulated private rental sector meets the needs of a dynamic economy: anyone arriving in London can find housing to suit their purse and preferences within a day: a penthouse or villa for the rich; a shared bed in a damp cellar for the poor.

Finally, the housing crisis has become a crisis for the productive economy as land used for industry, workshops and other economic activity can be sold at prices between three and ten times higher if it can be switched to speculative housing use. Planners, under strong pressure from politicians — and all of them bewitched by supply-side economists — have permitted and encouraged this switch, ignoring the erasure of economic life and useful services which had existed on this land.

In this context, there are the beginnings of resistance: untidy coalitions of housing tenants — always rather fragmented by the different kinds of landlords they confront — environmentalists, small and medium enterprises and neighbourhood associations. London has always had a tradition of micro-local activism and the challenge has been to knit local struggles with policy debates at city-wide scale. My own involvement has been with the Just Space network in which about hundred organizations support each other in this activity: building both organizational capacity and counter-narratives to the neoliberal orthodoxy.[2] This is the forging of new *communities of practice*, especially in the governance of landed commons: the streets, green (and blue) space and the social housing estates so hated and demonised by the elites.

In the present conjuncture, we have the national state pumping billions into radial transport infrastructure so that the growth can continue, fuelling land and property markets where the value is harvested by owners, investors and their attendant professions. The local state fosters densification on multiple fronts (though not in the most privileged areas) and prizes open new investment opportunities on former social housing and industrial sites. The central bank is aware that the financial system is at risk of this bubble bursting. We shall see. Meanwhile the challenge is to grow the critiques and resistance from the bottom up, maintaining exchanges with other scales and movements in other regions and countries.◄

---

2. The website http://justspace.org.uk has links to official and oppositional reports and academic work, together with campaign documents and plans. The network is part of the European Consortium for Rights to Housing and the City http://www.housingnotprofit.org/en and has links with INURA.org

06

Neighbor Helping Neighbor
GIVE NOLA ♥ LOVE

New Orleans © Angeliki Paidakaki

NEIGHBOURHOODS
AND COMMUNITIES

61

13. REFLECTING ON THE BOM: THE TRANSFORMATIVE POTENTIAL OF NEIGHBOURHOOD INITIATIVES

Patsy Healey and Jean Hillier[1]

Many planning and development scholars have been interested in the struggles through which marginalized citizens negotiate and resist oppressive governance practices which they encounter in their everyday lives. Integrated Area Development, developed by Moulaert et al. (2005) in the context of neighbourhood development initiatives focuses on the transformative potential of such initiatives, looking at the local as a more tangible, better and more just and democratic scale at which to organize change (González et al., 2010: 49).

The ALMOLIN framework — alternative models for local innovation — is an exploration of the potential of social innovation that was formulated by the EU-funded multi-case project on Social Innovation in Governance in (Local) Communities (SINGOCOM) (Moulaert et al., 2005). The term *social* carried the idea of a richly integrated concept of human life, in opposition to the dominant focus on economic returns and real-estate value. Drawing on experiences of the social movements of the 1960s and 1970s, Moulaert concluded that the *very local* of the neighbourhood was a significant site of social change (Moulaert 2010: 4-5). Initiatives were evaluated in terms of how they helped to satisfy human needs, expand resource access rights and enhance capabilities, and to change social relations, power structures and the modalities of governance (González et al., 2010: 54-55; Moulaert et al., 2005, 2010; Swyngedouw and Moulaert, 2010: 219). The work brought deep commitment to social justice with a spatialized sense of the dynamics of social movements, and analytical awareness of the multiscalar complexity of institutional dynamics.

These qualities are well displayed in the case-study, the *Buurt Ontwikkelings Maatschappij* (BOM) in Antwerp, a potentially transformative neighbourhood initiative. The BOM was established in 1990 as a public-civil society partnership between the municipality, the Flemish region and civil society actors (Van Hove, 2001). Working in marginalized

1. Acknowledgements: Jean Hillier's contribution to this chapter was written while she was based at Zhejiang University, Hangzhou, China. Jean thanks Zhejiang University for affording her this opportunity.

neighbourhoods of Northeast Antwerp, this local development agency pioneered citizen-focused initiatives addressing housing, work opportunities, knowledge enhancement and well-being in an integrated way. It expanded to other neighbourhoods in the 1990s and became a model for Flemish, Belgian and EU initiatives in urban neighbourhood development. For Moulaert and Christiaens, the fascination of the case lay in the way "social innovation interacts with the institutional dynamics of the places where it occurs" (2010: 175).

Through the BOM, residents discovered bottom-up opportunities for action in the pluralist political practices in Antwerp. Its approach centred on promoting projects, rather than developing processes — the focus of previous initiatives — along with partnership and increasingly active, project-specific participation from residents (Van Hove, 2001). Moulaert and Christiaens (2010: 178) summarize the BOM's multiple functions developed in this first phase as: pioneering, revealing needs, creating *impulse* through investing in the neighbourhood, acting as a catalyst promoting interaction among diverse groups, building opportunity structures for empowerment through new ways of co-operation and acting as an engine of change through promoting interdependent projects.

BOM thus created social networks and an action space for vulnerable groups (Moulaert and Christiaens, 2010: 176-177). However, as with so many neighbourhood-focused initiatives which flourished across Europe in the 1990s, BOM's engine ran out of transformative force in the 2000s. This was partly a problem of its own internal design, notably the dependence on public funding and its focus on individual, often competing, projects which pulled against the philosophy of integration. BOM also discovered that its approach was not easily replicable in other neighbourhoods. Other challenges emerged as shifts occurred in the wider institutional context (Christiaens et al., 2007). The City Council was riven with financial scandals, while a far-right opposition in the City and in Flanders (the *Vlaams Blok*) promoted a critique of non-profit agencies and the focus on social welfare generally. Public resources for neighbourhood-focused investment were reduced at municipal level and other grants came to their time-limits.

The problem for BOM was that it was too close to public agencies and too dependent on their finance. In the mid-2000s, some of BOM's key activities were subsumed into regular city services. The socially oriented approach pioneered by BOM gave way in favour of a real-estate driven urban regeneration policy driven by new public management principles, reflecting a market-oriented growth policy for the city as a whole (Christiaens et al., 2007). This seemed a "sad end of the story" (Moulaert and Christiaens, 2010: 181) at the time.

The BOM story, which resonates with so many others across Europe, could be interpreted as the failure of neighbourhood-based transformative initiatives to gain leverage against the creeping hegemony of neoliberalism. In contrast, Moulaert et al. (2010), in *Can Neighbourhoods Save the City?* regard local, socially innovative initiatives and activism as creating institutional sites which generate hopeful experiences of alternatives to the dispiriting decay of opportunity and ways of life which once sustained people. They emphasize, however, that developing such alternative possibilities is a slow process, requiring reliable sources of funding and a supportive governance context (Martinelli et al., 2010: 217-218). They also note that the relationship between civil society activism and formal government agencies is likely to be tense, often riven with suspicion and challenge. And when this tension gives way to a more co-operative and partnership modality, there is always the risk of "capture", as illustrated in the BOM case, by the priorities and practices of the larger formal agencies. As Swyngedouw and Moulaert (2010: 228) argue: "effective innovation initiatives involve a certain *distance from the state*, although they may exploit or, better, find fruitful accommodation within state institutional *folds* or protective *umbrellas* and enter into respectful collaboration".

How does this conclusion stand several years after the financial crisis of 2007/8 and the related *austerity* agenda pursued in many countries? While the corporate sector with its multi-national reach has largely sustained itself, and many have benefitted from the digital technology revolution, the scale of inequality between those who are flourishing and those on the downside of such changes has opened up alarmingly. Whereas in the 2000s, the challenge for several urban neighbourhoods was to resist the intrusion of large-scale speculative real-estate city-centre and waterfront projects, many have since stalled. Health, welfare, education, training, security and environmental services previously supplied by local governments have been contracted out to private agencies. These trends have fragmented service delivery and, for many people, disrupted their understanding of how to access what they need. In this context, the *tension* between state agencies and civil society seems to be taking a new turn.

In Europe, a new generation of neighbourhood and village-based practical activists has appeared, taking initiatives in health and social welfare provision, community energy projects, the management of public spaces, sustainable food production, affordable housing – sometimes in co-operative modes, co-operative banking, provision of leisure facilities, construction and transport services and development aid (Wagenaar et al., 2015: 559). Many of these small-scale civil society enterprises have arisen from local mobilization from within civil society, responding to gaps in provision as public services shrink, but also experimenting with new products, services and modalities of governance, as in the social innovation initiatives researched by Moulaert and colleagues. Always vulnerable, many have managed to grow and survive by energetic networking both among local residents and with state agencies and market organizations, skilfully riding the tension between "capture", co-operation and challenge as they seek the resources and support needed for objectives set locally. In contrast to many urban regeneration initiatives in the 1990s and 2000s which arrived with a funded programme and invited communities to participate (for the UK, see Bailey, 2012), these civil society initiatives intrude "uninvited" (Cornwall, 2008) into the surrounding governance landscape. Many are showing that public services can be delivered with quality and efficiency by small-scale, locally-rooted agencies, in contrast to larger more impersonal agencies operated by state and market providers. Although benefitting in many ways from state funding and technical support, those enterprises formed in, or surviving into, the 2010s, have had to be very aware of the need to build up their own resource base, through the ownership of assets, such as community buildings, premises for small firms, self-sustaining social enterprises, and the mobilization of volunteer labour. Several, like the BOM (Van Hove, 2001), have been inspired by the idea of *asset-based community development* emphasizing community assets rather than *needs*.

Some consider such initiatives as just a desperate response to the social crises created by the austerity programmes inspired by neoliberal ideas. But others researching the phenomenon (for example Healey, 2015b; Wagenaar and van der Heijden, 2015) come to similar conclusions as Moulaert and colleagues, noting the innovative and experimental orientation of many enterprises and their transformative potential. For example, in rural Northumberland, where the lead author has been actively involved, local activists are pushing forward new resident-centred and integrated ways of working into governance arenas of extremely severe financial cutbacks (Healey, 2015a, 2015b). In this context, the authority can no longer operate as primary service provider and deliverer as of old. It too is having to experiment, learning to respect what civil society agencies can deliver, and exploring how to provide the kinds of supportive institutional *folds* and *umbrellas* within which such initiatives can flourish.

Along with Moulaert and colleagues, and also those concerned with the expanding *commoning* movement (Dawney et al., 2016), many of us interested in these small-scale civil society enterprises are exploring a theory of social change and transformation which emphasizes the role of the *local*, understood as an institutional site where people not only care about their own life trajectories, but have also to share the public realm of their spatial worlds with others as they move around their daily spaces of living, these sites of localised collective action carry the potential to become learning arenas for what it takes to respond creatively to new challenges and to work with others with different views and experiences. Local actors generate considerable social capital, creating a *public sphere* while caring for what they share *in common* (Healey, 2018) . This enables both self-development and socio-economic and cultural cohesiveness, whilst facilitating valorization of the social ambitions of market agents and encouraging the creative power of civil servants and government agencies (Swyngedouw and Moulaert, 2010: 232-233).

There is, however, always the potential to be caught in a "localist trap" (Davoudi and Madanipour, 2015; Moulaert et al., 2005), where the self-selected impose unchallenged their view of the public realm and the modalities of the public sphere on others, or where there is a "blind faith" (Martinelli et al., 2010: 215) in the power of local agency to deal with local issues, ignoring the inherent multiscalar spatiality of development processes. Exclusionary, often inward-looking, communities rarely bring about real local empowerment. This is where Moulaert's message of the need for multiscalar engagement with such initiatives is so important. It remains essential that there is some form of collective oversight to care for both local social justice and environmental quality and also the relational impacts of what happens in one place on people in other places. Yet, as Moulaert et al. (2005: 1978-1979) suggest, this should not be a "Russian-dolls" model where local institutional dynamics conform completely to "higher" political decision-making and institutionalization. *Bottom-linking* (Garcia, 2006; Moulaert, 2010) is also important, in which small-scale enterprises benefit from networking with other agencies which share knowledge and support.

Moulaert et al. called for more academic attention to these kinds of initiatives. Developments since then underline that call, as old forms of provision have been undermined or been shown to be inappropriate. Some messages from their research, and later work on civil society enterprises, have already become clear. Firstly, it is more important for public policy to provide a supportive environment for such enterprises to grow than to arrive with a defined programme and "invite" depoliticized forms of participation (Cornwall, 2008). In addition, civil society activists need to learn how to ride the tension between autonomy and engagement with both public agencies and market actors and, in particular, work out how to build the capacity to survive beyond any particular public policy turn. In building that capacity, they will be faced with their own tensions between sustaining flexible, experimental and innovative capacity and over-institutionalizing themselves, so that they become too introverted to continue to explore their relevance and respond to new challenges. In turn, politicians, policy experts and officials should neither regard their expertise and authority as under threat (Healey, 2015b) nor react with attempts at control, but rather recognize that building local capacity and capabilities takes a long time and needs local roots, which may be continually changing. Short-term interventions pursued in ignorance of fine-grained local political dynamics may be as damaging as the withdrawal of a valued service. Finally, just because initiatives are local does not mean that they are accepted locally as inclusionary, legitimate and/or accountable. Legitimacy within a place-based community is an often fragile social accomplishment, different from simple "buy-in" to activities initiated externally. There will always be a tension between the risky anarchism of initial protest and experimentation, and the hard grind of building relations, which can draw and mould resources to make a difference in a particular locale. In this manner, the potential for achieving the key elements of ALMOLIN – enhancing capabilities to satisfy needs, to change social relations, power structures and the modalities of governance – may be facilitated.

The work on Integrated Area Development and social innovation is foundational for researchers interested in these and related issues. Over the past three decades Moulaert et al. have provided a coherent, theoretically-grounded focus on the potential of spatialized or area-based urban communities as institutional sites for progressive social transformation. They have insisted on viewing *progress* from a social, daily life perspective. They have also continually emphasized that such transformative initiative must always be understood in the context of the multi-scalar institutional context of state and market initiative as it plays out in particular situations through time. The work speaks to an expanding academic interest in progressive localism (Featherstone et al., 2012; Healey, 2015c; Williams et al., 2014) and also to both the struggles of activists working in city neighbourhoods and villages across Europe and those of progressive politicians, policy experts and officials seeking to evolve ways to refocus and rebuild our contemporary fragmented and often dystopian governance landscapes. ◀

14. SOCIAL INNOVATION AND COMMUNITY DEVELOPMENT: A PERSONAL REFLECTION

Diana MacCallum

DEFENDING SOCIAL INNOVATION – WHY DO I BOTHER?

I first engaged with research on social innovation through the European KATARSIS[1] project, which aimed to facilitate discussion about developing a coherent methodological framework for social innovation research. In KATARSIS the discussions were informed by a number of case studies which, in another time and place, I might have described broadly as *community development projects*: actions which built networks and collective capacity to address the needs of people whose interests and/or opportunities had been left behind by the advance of late capitalism. I'll be honest: it puzzled me that I, as someone – an Australian – who liked to think and read about community development practice and democracy in planning, had barely come across this discourse of "social innovation" whose aims and concerns overlapped so much with my interests.

Since then, things have changed. Social innovation is everywhere. In fact, *social innovator* has become a self-descriptor of choice for people – especially young people – who find ways of implementing bright ideas in response to social problems. In my country, a raft of organizations, businesses, public agencies and funding mechanisms have sprung up to support socially innovative projects, and corporate culture is increasingly centred on *hubs* that are said to facilitate innovation through social interaction, shared governance, and continuous exposure to evolving communication technologies.

This wide take up of the term, in an Anglophone country where policy at every level is steeped in neoliberal assumptions, speaks as much to its rhetorical appeal as to its success as a model of community development. In particular, it rides on the back of a general appeal to "innovation" – transcendent, decontextualized – as the solution to all problems and the saviour of the lurching national economy (thus, a central pillar of

---

1. KATARSIS (Growing Inequality and Social Innovation: Alternative Knowledge and Practice in Overcoming Social Exclusion in Europe) was a European Commission's funded Coordination Action under FP6 (2006-2009) http://katarsis.ncl.ac.uk/

Federal policy). I often hear *social innovation* described by people I respect as a malignant buzzword – a reframing of worthy causes in the language of entrepreneurship, emptying them of their emancipatory meaning. The discursive sleight of hand that allows *social innovation* to become a career choice also serves a technocratic, market-led approach to its institution in policy. Like *leadership* in the 1990s, a phenomenon that emerged from the practices and interactions of people with problems has become appropriated and reconceived as a field of generic expertise, one which supports the neoliberal project by creating new, marketable products to fill the gaps left by the withdrawal of the welfare state.

In this, Australia is quite typical. When I revisit the *International Handbook on Social Innovation* (Moulaert et al., 2013b), I am struck by how many of the chapters respond specifically to these concerns. The prevalent influence of a business/management version of social innovation on (especially) European policy, for instance through BEPA (2010) and the Young Foundation (Murray et al., 2010), was causing considerable disquiet. The book is saturated with calls to reassert the roots of social innovation, its spatio-temporal territoriality, and its grounding in the experience of excluded and marginalized people. In the years since that book's publication, the technocratic trend has only increased.

We are, no doubt, all familiar with organizations for which *innovation* has become a marketing slogan, and in which frequent calls for staff to "innovate" are thinly veiled code for "find ways to save money"; it's easy to become discouraged about the potential of innovation-based discourse to challenge injustice and exclusion. So, why do I still bother? Firstly, because innovation – social and otherwise – is important. We do need new ways of dealing with old and new problems, to paraphrase one of the "ready to wear" formulae beloved of policy makers (Australian Government, 2013). In the face of the environmental, ecological, economic, demographic and political crises that seem to be a constant feature of contemporary existence, we really cannot keep doing things the same, nor return to the alleged comfort of the past (notwithstanding the recent shocking successes of populist nationalism). Secondly, I defend it because I've seen it. We all have. People everywhere – social innovators, if we must – are doing amazing things which improve both lives and relations in their communities, at the same time challenging conventional market-driven and bureaucratic modes of service delivery and civic participation; the European projects SINGOCOM, KATARSIS and SOCIAL POLIS included many good examples, and we can all cite many others (Moulaert et al., 2010, 2013b).

So, I resent the conservative annexation of a word that can and should convey a mode of progressive change, and I see in the analysis of Moulaert et al. as something of an antidote. Underlying the proliferation of uses and meanings of *social innovation* (as strategy/practice/objective/outcome; as instrumental/designed/emergent; as specialized/extraordinary/everyday; as bottom-up/top-down/bottom-linked and so on) is a strong conceptual core which can enrich community development practice, firstly, by orienting it to social inclusion and justice and, secondly, by providing an analytical robustness grounded in contextual understanding. Social innovation is a reaction to social exclusion, by the excluded. It meets human needs neglected by the state-market apparatus, collectively empowers the marginalized, and leads to transformation in social relations. It is a territorial phenomenon, taking place in particular contexts and communities defined by social (and institutional, political, cultural, historical) associations rather than strictly by spatial scale.

ENRICH COMMUNITY DEVELOPMENT PRACTICE – HOW?

In KATARSIS we did not talk about *community development projects*, but *socially creative strategies* to combat exclusion. While this may seem a mere label (though, are labels ever "mere"?), I find the distinction helpful in that the latter term disambiguates several points that the former obscures. Firstly, it makes explicit the crucial role of creativity in producing social change; that is, it foregrounds the innovation aspect, and excludes the possibility of *template* approaches to community development. Secondly, it opens up the temporal and spatial dimensions, eliminating the misleading implication of self-containedness – clear beginning and end points, tightly-defined objectives, and set management procedures – that the word *project* can imply. Thirdly, and very importantly, it highlights the strategic agency of the excluded community itself, and reorients the impetus for their action from *development* – so often understood in predominantly economic terms – to exclusion. This also places the community's problems within the bigger picture of societal relations rather than originating in their own deficits, and highlights that sustainable solutions to problems of marginalization lie in social change, rather than local fixes. The connection between *socially creative strategies* and social innovation is, by implication, a rendering of micro-macro relations, recognizing that the things we do on the ground are embedded within, respond to, and sometimes transform the broader social and political environment.

I do not mean at all to suggest that community development practitioners see their work as taking place in a vacuum, or without the participation and empowerment of communities. In my experience, most have been acutely aware of (and very articulate about) the big-picture political-economic histories that shape community needs and potentials, and fully respectful of – often a member of – the community they work with. Moreover, recognition of the dangers of deficit-oriented community development has shaped an important set of practices, collectively referred to as assets-based/strengths-based community development (ABCD, ASBCD – see for example Kunnen et al., 2013; Gibson-Graham and Roelvinck, 2009) that actively work to begin from a point of strength and empowerment. The point of the above example is not to criticize those involved in the relevant professions, but to illustrate how the conceptual framework of social innovation shapes our analytical and strategic attention, and to argue that this has concrete implications for practice.

To tease this out a little, the social innovation framework disrupts and renders unsustainable the dichotomy between bottom-up and top-down development – thus, the coining of *bottom-linked* as a descriptor for successful SI strategies (Eizaguirre et al., 2012). At the centre of SI is not one thing or the other, but the complex, ideologically and logistically fraught interaction between people, institutions and society. As such, perhaps the proper stuff of SI-informed practice is the creation – or recognition – of spaces of opportunity to positively reconfigure the dynamics of this interaction. Achieving this may not, in fact, involve the formation of an authoritative account of what the problems are, shared objectives, a catalogue of resources, an overall strategy, agreed methods for monitoring success – or any such conventional planning technologies. Indeed, it has long been recognized that the whole idea of consensus on such issues may be intrinsically exclusionary (Hillier, 2003). Should people accept, for example, the reality of anthropogenic climate change, the beauty of street art, the personhood of animals, the desirability of multiculturalism, the secularity of the state (I'm making even myself feel uncomfortable) and so on in order to participate in building community resilience and wellbeing? Really?

It seems to me that the types of spaces in which SI emerges are often constituted in collective action at the boundaries between institutions – between business practice and the social economy; State service delivery and Indigenous Law; education and industrial relations; arts and food production; recreation, media and environmental management; and so on and so forth. These are spaces of co-learning, in which communities of (often hybrid) practice may be born, with shared language, knowledge, sense of purpose, norms, relations... (Wenger, 1998). Insofar as these communities of practice are more inclusive, more equitable, and permit participants to have higher expectations of the world and of themselves, we can say that they represent social innovations. Sustainability and societal leverage, though, depend on institutionalization – reification, in communities of practice terms, of new practices – especially in ways that can speak to power. The local creation of an employment opportunity, delivery of a service, construction of a facility, establishment of an inclusive decision-making process – all very good and potentially empowering things – are not the end point, so to speak, of social innovation. Reification of practices need not necessarily mean the creation of *hard* policy or infrastructure; it can take many, often softer forms, including embodiment in the habitus of participants, redesigned local governance arrangements, political representation and/or acknowledgement, publication of norms, and so on. While not always easy (or even possible) to achieve, it seems to be an important consideration not only for analysts, but also for practitioners; it is a crucial one if we are to defend the transformative potential of SI.

At one of the later KATARSIS conferences, I and about twenty others participated in a discussion session on how our academic work could contribute to practice. We found it surprisingly difficult to focus on this question and kept slipping comfortably into talking about how observing and participating in practice contributed to our scholarship. In part, this reflected a recognition that, in the KATARSIS case studies at least, the practitioners knew what they were doing; they already demonstrated remarkable analytical insight, practical judgement, and social, administrative and manual skills. Thinking back on it, I wonder if it also reflected an institutional culture in which we are – after thirty or so years of economic rationalism – accustomed to critical scholarship speaking beautifully to itself, but generally being sidelined when it comes to mainstream influence. In retrospect, given the significant role of academia in disseminating and shaping ideas – in creating the imaginaries that become objects of governance (Sum and Jessop, 2013) – I do not think we were trying hard enough.

Since that conference, a commitment to transdisciplinarity – for example through partnerships that involve actors beyond the academy in problematization, research design, analysis, dissemination and end use – has become a central tenet, or at least a strong recommendation, of SI scholarship. This is a good thing. Indeed, participatory action research, with its very similar commitments (Macguire, 1987; Kemmis and McTaggart, 2000), has long been a privileged methodology for research in community development, particularly in ABCD. The two traditions surely still have much to talk about (Kunnen et al., 2013)!

SI-INFORMED GOVERNANCE?

Tom Montgomery (2016) recently characterized the SI field as consisting of two broad and competing (rather than commensurate) paradigms, the *technocratic* (exemplified by SI as a management discourse) and the *democratic* (including the work of Moulaert and colleagues since the 1980s). When we look at the term's uptake in government and corporate policy and programs, it is hard not to think that the former paradigm is winning. But, as we know well, formal policy is not the full story.

It's not even the State's full story. Some of the most socially creative people I know are public servants, street-level bureaucrats who find spaces within and ways through adopted policies and programs to enact daily resistance both to social exclusion on the one hand, and, on the other, to the influence of discourses which blame the poor for their condition. As an example, Western Australia's 'Planning for Aboriginal Communities' program (WA DoP n.d.) – despite its bland public face (which serves a useful purpose) — has just such a story behind it. Less dramatically, we can cite all those people making sensitive everyday decisions about public housing and welfare (for example see Marston, 2004). I also know people who have created new business models that allow them and their associates to live and work largely independently of drive for economic growth (see Houston et al., 2016). I know people who spend their evenings making music that expresses the experience and roots of deprivation; and I know other people who spend every weekend helping them build the equipment they need to do that.

These are all people who sustain a strong sense of political purpose, in spite of (and often in direct response to) the increasingly managerialist *post-political* discourse that represents the institutions in which they work (Wilson and Swyngedouw, 2014). Their actions take creative imagination and, to varying degrees, meet needs, build social resilience, and reshape public attention. *Social innovation* — in Moulaert et al.'s sense — is a good descriptor for what they do. And that is why I continue to defend the term against accusations of emptiness — because it not only expresses something that I can see and wish to honour; it also can help literally convey that reality — a kind of progressive practice — into other governance arenas. ◄

15. ÉVORA

◆

Patricia Rego and Isabel André

This contribution presents and reflects on the local platform *plat.for.evora* which promotes territorial cohesion and social innovation through cooperation and exchange between third sector organizations in Évora – the capital of Alentejo region, Portugal – and its surroundings. This platform was developed in collaboration with local actors and the University of Évora in 2014, following a study about the role of the third sector for social innovation in Évora (André et al., 2014). The study, conducted in collaboration with Eugénio de Almeida Foundation (FEA) and other local actors and researchers, emphasized the role of civil society organizations in promoting the values of solidarity, reciprocity and collaboration to achieve greater equity for all.

The main objective of *plat.for.evora* is to promote the (re)construction of proximity social networks with the active participation of local entities as community mediators. The improvement of *proximity* networking, through resource-sharing between organizations allows reinforcing institutional relational capital. These aspects emphasize the dimension of social innovation as innovation in social relations (MacCallum et al., 2009; Moulaert et al., 2013b) opening doors to the reinvigoration of the community. In this respect, it is important to understand how the networking proximity contributes to the progress of communities, the role of leadership and the active involvement of organizations and citizens and, finally, the added value of the intersection of various types of knowledge (practical and theoretical) and its relevance to local progress and social cohesion.

Collaboration is at the core of *plat.for.evora*, echoing Kamensky and Burlin (2004) who stated that 'collaboration occurs when people from different organizations produce something together through joint effort, (shared) resources and (co-)decision making, and share ownership of the final product or service'. The development of *plat.for.evora* followed this collaboration process through the following steps:

1. Analysis of the territory (the municipality of Évora) through statistical profiling, diagnosis (SWOT and *Asset-Based Community Development* (ABCD) Analysis), and knowledge of the existing networks (third sector organizations).

2. Reflections that build a common vision for the territory based on the debate among third sector organizations. Critical thinking and social innovation, especially the transformation of social relations, are the key elements of this reflection process.

3. Deliberations to understand and enhance the links between organizations and other entities and experts to build a territorial vision. The relationship between these actors is formed using different approaches (workshops, conferences, forums, laboratory of ideas, and so on). Neighbourhoods and their sense of community are crucial. The goal is to identify the collective interest and common good.

4. Capacity building activities and co-production of knowledge. Here we recognize the importance of good communication, enabling discourses and media (workshops, training).

5. Experimentation actions to demonstrate and promote organizations' dynamics and social innovation (experience as a learning process, pilot-experiences and living labs).

6. Mediation for the identification of adequate vehicles and networks to stimulate a common/shared vision of the territory.

The methodological path developed in *plat.for.evora* included quantitative tools – surveys about the platform results addressed to local actors and network analysis (based on formal and informal ties between local actors) and qualitative tools – highlighting the results of focus group sessions and the outcomes of creative ideas labs.

The critical thinking debate sessions and experimental social innovation initiatives were crucial for the positive first evaluation of this local platform. In this context, a debate was organized on the initial results of the initiative in the Workshop on *Socialising Living Labs* at KU Leuven in March 2016. First year results of the platform showed that forty-seven organizations participated in plat.for.evora activities, of which thirteen were from outside the municipality of Évora and thirty-four were from within Évora. Of these thirty-four, just two are located in rural areas *(freguesias)*. The number of participants in different activities is diverse.

The organizations participated in one or more actions/activities that can be classified as reflection encounters, capacity building activities, field trips to visit and discuss positive experiences of social innovation, and labs to stimulate and support new social ideas or projects.

The first results of *plat.for.evora* indicate the relevance of *creative ideas labs*, which is an initiative focused on the tools, knowledge and networks needed by organizations to implement local management processes, to be sustainable and to add value to the existing resources. More significant results, particularly in sharing these experiences of new forms of governance, need more time because these are long-term processes.◀

Photo: Évora © Patricia Rego

07

SOCIAL INCLUSION THROUGH CULTURE AND ARTS

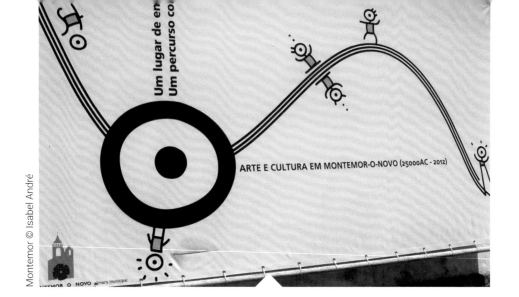

Montemor © Isabel André

16. INSPIRATION AND EMOTIONS: CULTURE AND ARTS ENGENDERING NEW URBAN PLACES

Isabel André[1]

Adversity and opportunity drive change and transform cities, generating disquiet, fermenting imagination and sparking challenges. They are necessary conditions, but they are not sufficient to generate socio-territorial innovation (Klein and Harrisson, 2006; Moulaert et al., 2013b; Fontan et al., 2005), understood as the expression of desired futures in encounters between visions, between the impossible and possible, and between confluent or discordant practices (Pinder, 2015) which are, nonetheless, united in seeking to build a more just, inclusive and democratic city.

Communication, especially in regard to the expression of unmet needs and dialogue between actors, is essential in terms of generating innovative and feasible solutions. In this precise connection, the arts are revealed as important vehicles for building the cities of the future. Artistic expression allows us, on the one hand, to symbolically identify undesirable or unfeasible present situations and, on the other, to anticipate the future by

---

1. This is an unfished paper by Isabel André who passed away before its completion. Diana MacCallum and the editors reworked the document to its present form.

suggesting new answers to unmet needs (Smiers, 2003). The use of metaphor, in all its forms, enables us to transcend the obvious and enter into the realm of emotions – it allows communication beyond the common language (Ruby, 2002; André and Abreu, 2009; Ley, 2003).

Focusing attention on the contemporary city – ruled in the interests of neoliberals yet at the same time confronted by protests and actions presenting alternative approaches – it seems that imagining the urban dynamics of the future involves acknowledging a greater role for culture and the arts, whether from the perspective of the market and real estate capital or from one committed to progressive social change.

In fact, divergent urban visions expressed through the arts have the ability to distinguish cities, making them "unique places", and also, contributing to their symbolic value (associated with belonging and identity, as well as external recognizability) and capacity to celebrate. That is, there appear to be significant confluences in the relationship between the city, the arts and culture. However, the intentions, objectives and concrete results are often clearly discordant with a vision of urban justice, reflecting a standardized *creative city* in the vein of Landry (2000) and Florida (2002) – a conception fully immersed in, and adopted by, neoliberalism.

The discovery of *the soul of neighbourhoods and cities* – through the support and participation of artists – allows us to believe that *another (creative) city is possible!* With this in mind, this text reflects on the arts and artists as crucial actors in the transformation of urban places, by considering their mutual gains: what do the arts and artists give to the city? And what does the city give to the arts and artists? (André et al., 2013; Le Floc'h, 2015). This *exchange* can be understood as the impetus for and also the outcome of cultural scenes: physical and virtual spaces where inspirations are stimulated and emotions bubble.

WHAT DO THE ARTS AND ARTISTS GIVE TO THE CITY?

From the late nineteenth century, and more profoundly since the 1960s, art has come out of the elite institutions and into the streets, often expressing the fight for social justice. The academies, galleries, theatres or museums opened their doors and, in addition, the artistic production and dissemination started to happen in many other places, in everyday spaces of ordinary citizens.

Public artistic expressions – using different languages or mixing them – reflect societal tensions and individual and collective disputes (André et al., 2009) revealing unsolved problems, giving voice to conditions of adversity and expectations of advancement. Through the use of metaphor and subjectivity, the arts connect elements that were separated, and this ability has been recognized as a crucial element of the production of innovative urban space. In this same line of thinking, the arts can transform indignation and outrage into creative force.

"As we approach the main centre of psychiatric health in the city of Lisbon, a vast area of 22 hectares in a central location in the city, and walk along its outer wall, we can see it is covered in paintings by street artists representing various expressions of mental disease. In fact, we are dealing with an intervention that reveals severe social tensions, contributing mainly to mitigate the boundary between normality and disease and to fight the strong stigma associated with mental illness."

"The movement of the 'indignados' that emerged in 2011 in Spain as a reaction to neoliberal austerity policies has mobilized artistic expressions that, with the energy of protest, give rise to new public spaces solutions, literary works, movies, murals, musical pieces, among many other artistic works. According to Spanish magazine Extracine (Álvarez, 2013), between 2011 and 2013, 20 movies were produced on the 'indignados.'

Promoting recognition and distinction, urban art – visual or performative – can increase the self-esteem and confidence of local collectives, fostering creativity in various spheres of urban activity and facilitating the emergence of new solutions to problems. At the same time, it enables the re-making of spaces, lending light and brightness to otherwise invisible urban dynamics (Vahtrapuu, 2013) associated with freedom, belonging or the right to the city.

Bringing these various contributions of culture and the arts to the production of an alternative city together, we can feel inspired by the way they create a sense of transcendence and anticipation – the essence of a real utopia and the seed of the society of tomorrow (André et al., 2013). In this way, the arts are often associated with social innovation and the regenerative capacity of places (Capel, 1996; Moulaert et al., 2004).

From the approach of the arts to the city – through the artists, cultural actors and urban activists – cultural scenes can emerge mainly in public squares, in areas related to transport (for example old stations and warehouses), in former industrial spaces, gardens, and so on. Such scenes – outside the boundaries of corporations – unlock skills and combine many talents. Scenography – the articulation of elements in creating scenes – becomes central to allowing the expression of the diverse artistic languages in a given territory. It also contextualizes artistic actions that thereby lose pretensions of "neutrality" and gain real capacity to deal with urban problems, for instance through conciliation between public authorities, the private sector and the third sector.

"A few years ago, a renowned Portuguese choreographer (Rui Horta), told me that when he wants to have new ideas and 'give wings' to his imagination, he goes to Lisbon, Berlin or New York. The creation phase is passed in a quiet place in a small rural town (where the repetition of daily routines helps maintain harmony and balance, necessary for production). In the final phase of the work – diffusion – he returns again to the big cities."

WHAT DOES THE CITY GIVE TO THE ARTS AND ARTISTS?

The city also has important attributes for culture and the arts. It is a territory, with a unique context and historical path, although often disregarded, the city's geographical and temporal boundaries are important to give texture to artistic actions. According to Sack (1992), the territory is established by the borders between *ourselves* and the *others*, between being together or separated and, from a more abstract point of view, between the general and the particular. The territory permits or prevents the reinforcement of such barriers depending on nature, cultural meanings and social relations. At the same time, the territory contextualizes visions and actions, giving them a sense that joins the various parts of the whole. However, neither the frontiers nor the context exist outside of the territory's historical path. The passage of time may have left physical or intangible marks, but certainly builds an identity in which the memory and the expected futures can be reconciled or brought into confrontation.

The three aforementioned elements – boundaries, context and identity – make a space become a place (Tuan, 1977) and the city a network of numerous places – crucial conditions for artistic expressions. The urban territory, with its pressures, contradictions and tensions is a challenge for the artists – especially in relation to competing imagery and debates, both of which fertilize their talent. In the city, its places are associated with both the ephemeral and the stable, therefore unforeseen incidents occur and it is precisely these paradoxical links that seem to provide energy to the arts, in such forms as debate, production and diffusion.

Besides the sense that places give to the arts, cities increasingly provide institutional support to enable artists to overcome difficulties and challenges. Among the most relevant institutions are artistic schools, local authorities, museums, theatres, galleries and cultural centres. This assistance and services network is so useful to the artist, acting almost as a safety net. The city also gives artists new social and cultural environments in a wide web of relationships that allows them not only to expand and renew the dialogue but also to develop new plans and engage in various projects, thus expanding their range of opportunities.

The preceding paragraphs constitute a very optimistic discourse which present the city as almost a paradise for artists. The reality is not so!

Cities and those who govern them are very aware of the economic value of the arts. An urban space where artistic expression – usually that of elites – is present has real added value. In addition, it attracts tourists for whom cultural capital becomes a principal reason for travelling. The regeneration of urban spaces which has helped artists in many cities, in some cases naively and in others opportunistically, also often leads to the violent phenomena of social exclusion.

The mutual synergies between the city and the arts are played out in cultural scenes, through which the role of arts and artists in the emergence of new urbanities (Borja, 2011) is realized. Through such scenes conflicts and crisis can be reshaped in socially innovative ways.

CULTURAL SCENES ENGENDERING NEW PLACES

Cultural scenes are spaces with strong vitality, mobilizing energies in various directions and allowing the production of new urban *nests*. Scenography transforms spaces (frequently those which have been abandoned and decayed) into places with the ascription of new meanings, and insurgent and alternative places, in which protest and creative solutions thrive and contribute towards more democratic and just cities for tomorrow. These new places often emerge in periods of crisis when artists take to the streets, giving conflicts a voice and working to make Utopia a reality.

Although easily identified on the ground, cultural scenes are difficult to define, a complex concept in which various actors and agents are intertwined, with different scales and timings. As noted by Straw (2004: 413) 'the challenge for research is that of acknowledging the elusive, ephemeral character of scenes while recognizing their productive, even functional, role within urban life'. Cultural scenes can be organic or planned spaces, they may be informal or connected to business activity, insurgent or relatively disciplined, arising as 'spaces of artistic creation, as specialized urban centres arising as often from economic constraints and political strategy and from individual modes of appropriation of urban spaces'[2] (Boichot, 2013: 19). They act to mobilize artistic and cultural resources based on proximity relationships between the actors and stakeholders (Stern and Seifert, 2010) such as artists, citizens, associations, companies or public entities.

More than this, though, scenes are sites of insurgence – both aesthetic and socio-territorial. This insurgence is manifest 'through the reclaiming of invested spaces and the putting into place projects which break from classic modes of production and dissemination of art, mobilizing interdisciplinary artistic collaboration,

---

2. Translated from original text: 'espaces de la création artistique comme des centralités urbaines spécialisées résultant autant de contraintes économiques, de stratégies politiques que de modes d'appropriation individuelle des espaces urbains.'

and work studies transformed into places of demonstration'[3] (Aubouin and Coblence, 2013: 94-95). This interdisciplinary mixing means that the projects can take a hybrid and highly flexible nature, allowing artists 'to break away from classical institutional frameworks, and to dissolve traditional barriers between disciplines (live spectacle, material arts), between fields (cultural, social and educational), between amateur and professional practices, between the public sphere and the population.'[4] (Aubouin and Coblence, 2013: 94-95).

AND HOW DO SUCH CULTURAL SCENES CHALLENGE AND TRANSFORM THE CITY?

Cultural scenes prefigure new ways of organizing and producing urban space: 'scenes may be seen as ways of *processing* the abundance of artefacts and spaces which sediment within cities over time' (Straw, 2004: 416). The socio-territorial innovation associated with the production of new forms of urbanism involves changes in the field of the city's symbolic values but can also bring into being new social relations – based on proximity and the importance of collectives – and more democratic political models, which are socially and ecologically focused. The scenes connect network space (relational) with the space of places (identity). They also combine intimate, parochial and collective spaces. Through the artistic expressions they host, scenes contribute towards envisioning desired futures of cities as well as channelling discontent and protest, facilitating dialogue between urban actors with divergent interests, and allowing for commitments that resolve or mitigate potentially violent tensions. Moreover, artists have the ability to provide the city with new images and representations of itself and its relation to the world, stimulating the collective imagination, creativity and urban innovation.

Political, social and economic crises have a strong impact on the field of cultural values and practices, and particularly in the production and dissemination of art. If, on the one hand, crisis triggers a reduction in domestic consumption and public support for culture and the arts, on the other hand, it stimulates the emergence of creative solutions that meet the restrictions generated by austerity policies and economic downturn, and also encourages the formation of collectives where cooperation and networking emerge as new forces. This picture generated by the crisis is clearly felt in the cultural and artistic milieus and may be considered a kind of *fermentation* for the emergence of cultural scenes and, ultimately, the transformation of urban space.◄

---

3. Translated from original text: 'dans la revalorization des lieux investis et la mise en place de projets en rupture avec les modes classiques de production et de diffusion de l'art, par des collaborations artistiques interdisciplinaires, des ateliers transformés en lieu de démonstration'.
4. Translated from original text: 'de rompre avec les cadres institutionnels classiques, de faire disparaître les frontières établies entre disciplines (spectacle vivant, arts plastiques...), entre champs (culturel, social et éducatif), entre pratiques amateurs et professionnelles, entre public et population'.

Montemor © Isabel André

Montemor © Isabel André

Photo : Creating a mural together in public space – students of
Escola Básica 2, 3 do Alto do Lumiar © Ruth Segers

17. ACTION RESEARCH WITH AN ARTISTIC TURN: THE "WELCOME (W)ALL-MURO" PROJECT FOR YOUTH IN BAIRRO DA CRUZ VERMELHA, LISBON

◆

Ruth Segers

Welcome (W)all-MURO is a project to make a mural together with local youth in the public space of Bairro da Cruz Vermelha in Lumiar, Lisbon. Bairro da Cruz Vermelha is a fairly new neighbourhood located in the extreme north of the capital city, adjacent to the national airport. In the 1970s, the Lisbon Council built social housing apartments in this area to rehouse domestic and international migrants who had arrived in the 1960s. Migrants had put together a new homeland from *wood and zinc.* Today the slums have been fully removed and more social and private housing apartments have been erected as part of the social mix agenda of the Urbanisation Plan of Alto do Lumiar (PUAL) that started in the late 1990s.

The research aim of Welcome (W)all-MURO was to take the place related needs, attachments and personal capacities of young people into account in the co-creation process of making a mural, and to see what effect this process has on interpersonal cooperation.

Methodology-wise, the Welcome (W)all-MURO was a collaborative, situation based, and context-specific project undertaken by people with a common purpose: to provide empowering opportunities for learning and socialization of youth. The individuals involved represented the local institutions Centro Social da Musgueira, Escola Básica 2+3 do Alto do Lumiar, Junta de Freguesia do Lumiar and two universities: The Planning & Development Research Unit of the KU Leuven (Belgium) and the Institute of Geography and Spatial Planning of the University of Lisbon.

The mural was finished in April 2017 and depicts results of three different workshops with differently aged children and youth.

1. The mandalas are personal work by Mario, Claudia, Rodrigo and Joelma (15-19 years).

2. The "touching" spots in the neighbourhood are based on the photos of Miguel, Bruno, Guilherme and Marta (10-14 years). The spots were captured during a walking interview with fifteen young people through the neighbourhood and were translated into paintings.

3. A representation of the first social apartment buildings of the neighbourhood was the topic of a workshop with eight children (8-12 years). ◄

Creating a mural together in public space – students of Escola Básica 2, 3 do Alto do Lumiar

The Welcome (W)all–MURO at the pedestrian entrance of Bairro da Cruz Vermelha, LUMIAR, Lisbon

Social Innovation as Political Transformation. Thoughts for a Better World

© ndvr

08

SOLIDARITY AS GOVERNANCE

18. PUTTING SOLIDARITY IN ITS PLACE IN METAGOVERNANCE

◆

Bob Jessop and Ngai-Ling Sum

Governance and, more recently, metagovernance are notions whose time has come. Although the notion of governance has a long history, it was revived in the late 1970s to denote an alleged *turn from government to governance* in politics and analogous shifts in other societal spheres. This revival was followed from the mid-1990s by growing theoretical and practical interest in metagovernance (for a comprehensive review of the relevant theoretical and policy literature, see Meuleman, 2008). This chapter scopes the revival of metagovernance in the territorial and other spatial dimensions of regulation, including their implications for institutional and collective action dynamics, and socio-economic development trajectories. It particularly looks at the works highlighting the importance of solidarity as a critical but neglected principle or type of governance, especially in relation to the social economy, social innovation, social cohesion, community development, and the empowerment of networks for democratic planning. In short, taking solidarity seriously is essential in defining and constituting the social economy, its agents, and its dynamics and vital for a holistic approach to (meta-)governance (see especially Moulaert and Nussbaumer, 2005), which explains the title of our contribution. The emphasis on putting solidarity in its place in (meta)governance, is a distinctive contribution to the theorization and investigation of the aforementioned themes. It adds an explicit normative dimension to these debates, which are too often primarily instrumental, technocratic, or responsive to top-down government agendas.

EXPLORING MODES OF GOVERNANCE

Governance is a polyvalent and, too often, chaotic concept. This reflects the relatively *pre-theoretical* and eclectic nature of work in this area, the diversity of more rigorous theoretical approaches to governance, the heterogeneity of its subjects and objects and, a fortiori, those of metagovernance, and, of course, the different political traditions and currents interested in governance practices. Frank Moulaert has long confronted the first three challenges thanks to his interest in the potential of diverse theoretical approaches, even seemingly incommensurable ones, in illuminating the many topics on which he works (see, notably, joint synthesizing contributions to four European

Commission Framework projects SINGOCOM, KATARSIS, DEMOLOGOS and SOCIALPOLIS). Among the traditions studied in these projects are economic sociology, institutional economics, the German Historical School, world systems theory, political ecology, regional theories, regulation theories, and spatial planning theory. Thankfully, but unsurprisingly, Moulaert is far less tolerant of acquisitive, authoritarian, anti-democratic and inherently unjust uses of governance mechanisms!

In broad terms, governance refers to mechanisms and strategies of coordination adopted in the face of complex reciprocal interdependence among the actions, activities, and operations of autonomous actors, organizations, and functional systems. Three main kinds of governance are usually considered: the invisible hand of the market (exchange), top-down management (command), and reflexive deliberation among equals with different but complementary interests (networking). In this field, governance is sometimes identified only with the third mode (Moulaert and Sekia, 2003). This is clearest in its guise of reflexive self-organization and/or empowered networking based on continuing dialogue and resource-sharing among independent actors who aim to develop mutually beneficial joint projects and manage the contradictions and dilemmas inevitable in such situations. As Moulaert and Sekia (2003) and Moulaert and Cabaret (2006) note, however, networking can also work in very asymmetrical and exclusionary ways. A fourth mode of governance is sometimes explored too – but far less often than markets, imperative coordination, and networks. This is solidarity or, more specifically, unconditional solidarities based on identification with a (real or imagined) community. A crucial aspect of Moulaert's work is his concern with solidarity and its institutional and substantive adequacy to the challenges of promoting social innovation, creating and consolidating the social economy, securing social cohesion as a social policy objective, and managing local and regional economies for a resilient participatory democracy (for example, Novy et al., 2012).

Some governance theorists have correlated these types of governance to different sets of social relations, respectively: exchange with markets, hierarchy with the world of states and inter-state relations, self-reflexive governance with networks and society, solidarity with real or imagined communities. They also link them to different kinds of social logic. Thus, markets involve ex post coordination based on the formally rational pursuit of self-interest by individual agents; command corresponds to various forms of ex ante imperative coordination concerned with the pursuit of substantive goals established from above; and networks are suited for systems (non-political as well as political) that are resistant to top-down internal management and/or direct external control and that co-evolve with other (complex) sets of social relations with which their various decisions, operations, and aims are reciprocally interdependent. Frank Moulaert adds that solidarity is especially well-suited as the primary governance mechanism for the social economy. Indeed, Moulaert and Nussbaumer (2005) emphasize that the functional logics of markets and imperative coordination do not, and could never, deliver the results needed for a sustainable social economy. Only governance based on co-operation might achieve this.

Nonetheless Moulaert does acknowledge the necessary role of markets, hierarchy, and networks within a social economy context where they operate in the shadow of solidarity. It is important to *get the balance right* between these four forms. This challenging task is now known as metagovernance. The latter has been defined as, inter alia, the organization of self-organization, the regulation of self-regulation, the steering of self-steering, and as collibration (Dunsire, 1996; Jessop, 2011). While Moulaert does not refer to it in these terms, the concept (as opposed to the word) is crucial to his analysis and advocacy of the social economy. Seen in these terms, we could add that Moulaert's approach to governance is more society-centred (embracing economic, societal, and community-based variants) than state-centred in character. This also reflects his greater interest in local, urban, and regional government and governance arrangements over the form and activities of the national territorial state.

REGULATION AND GOVERNANCE

The regulation approach highlights the role of *institutional dynamics* in coping with contradictions and crisis-tendencies. Moulaert's early contributions to this approach already indicated an interest in governance, especially regarding the spatiality of regulation and the regulation (or governance) of socio-spatial relations (for example, Moulaert et al., 1988). The same concern is shown in his subsequent regulationist work (for example, Moulaert, 1996; Moulaert et al., 2007b). Regulationists argue that market exchange alone cannot secure economic growth or social stability because markets are inherently prone to market failure. They also note how markets and market failures vary across epochs, economic periods, and economic sectors. For these reasons, they examine the state's role in providing many of the extra-economic supports – material, institutional, policy-driven, and discursive – that enable markets to operate and/or that compensate for their inevitable failures. They do not interpret this in technicist terms but in terms of successive patterns of institutionalized compromise that shape state forms and activities. Equally significant are other extra-economic forms that help to stabilize capital accumulation – albeit unevenly and provisionally. Here regulationists discuss the role of networks, inter-firm linkages, norms, values, conventions, and other social forces in regularizing accumulation. These are also central to Moulaert's earlier work on territorial innovation and, latterly, on the social economy and social innovation and their regulation-cum-governance.

Different modes of governance not only have their distinctive functional logics but also their characteristic forms of failure, related to their differential privileging of some actors and interests over others, their specific forms of conflict, segmentation, and exclusion, and their respective forms of social resistance and counter-movements. It therefore matters how – by which institutions – governance, regularization, and normalization interact and how they relate to rupture, disintegration, and conflict and to renewed convergence and social integration (cf. Jessop and Swyngedouw, 2005). This highlights the importance of the differential articulation of diverse modes, sites and scales of economic and social organization that go well beyond market forces to secure the

relative coherence and coordination of different modes of governance. This is reflected in a provisional definition of socio-economic governance as the social relations that govern the functional organization of a (socio)economy or some of its components (Moulaert and Nussbaumer, 2005). A holistic analysis reveals close links between institutional innovation (especially social innovation) and modes of socio-economic governance oriented to cultural emancipation, community empowerment and consolidation of the local (social) economy.

While regulationists tend to focus on basic structural features of capitalism and their medium- to long-term constitution and stabilization, theorists of governance tend to focus on institutions and practices across many different social fields. Nonetheless, there has been a partial rapprochement between regulationist work and studies of economic governance at the sectoral, local, regional, national, and international levels. Whereas regulationists have studied different mechanisms of governance and their role in regularizing the key structural forms of the economy in its inclusive sense, students of governance have explored why different economic sectors have different modes of coordination, the problems of economic governance at different scales from the local to the global, the shift from government to governance in the state and inter-state systems, and the rise of networked forms of sociality and network societies. Based on such convergences, there has been a rapprochement among regulation theory, evolutionary economics, economic sociology, and, one might add, political science and international relations insofar as regulation theory provides appropriate vocabulary for other institutional and socio-economic approaches. In any case, a theory and definition of governance in new institutionalism encompasses many more dimensions than in new institutional economics (Moulaert and Nussbaumer, 2005).

GOVERNING TERRITORIES

Governance is related to debates on the territorial state in two main ways. First, states are hierarchical sovereign institutions, but inter-state relations are anarchic and in need of non-hierarchical forms of governance; and, second, constitutionalized violence is a last resort in democratic regimes and governments rely on other means to advance state projects and win legitimacy. This is even more important where social relations are not confined within national territorial boundaries but have variable territorial geometries, display tangled scales of social organization, and involve many networks.

Multilevel governance (Moulaert also writes of multiscalar governance) involves the institutionalization of reflexive self-organization among multiple stakeholders across several scales of state territorial organization. This has two implications. First, state actors would cooperate as negotiating partners in a complex network, pooling their sovereign authority and other distinctive capacities to help realize collectively agreed aims and objectives on behalf of all members of the network and its capacity to collaborate. They would operate at best as primus inter pares in a complex and heterogeneous network rather than as immediate holders of sovereign authority in a single hierarchical command structure. Thus, the formal sovereignty of states is better seen as one symbolic and/or material resource among others rather than as the dominant resource. Indeed, from a multilevel governance perspective, sovereignty is better interpreted as a series of specific state capacities (for example, legislative, fiscal, coercive, or other state powers) rather than as one overarching and defining feature of the state. So, states will supply other resources, too, that are not directly tied to their sovereign control over a national territory with its monopoly of organized coercion, its control over the national money, and its monopoly over taxation. State involvement is therefore less hierarchical, less centralized, and less directive in character. Other stakeholders contribute other symbolic and/or material resources (for example, private money, legitimacy, information, expertise, organizational capacities, or power of numbers) to advance collectively agreed aims and objectives. Second, in contrast to the clear hierarchy of territorial powers associated in theory with the sovereign state, multilevel governance typically involves tangled hierarchies and complex interdependence (see Jessop, 2016).

Moulaert's work reflects interest in regulatory and governance arrangements, or *government styles* (Moulaert and Nussbaumer, 2005), as the changing condensation and expression of (class) social forces and power relations. Regulation-cum-governance arrangements are regarded as mediating/disciplining mechanisms among tensions and conflicts and as oriented to ensuring the most functional/least disruptive reproduction system. This said, Moulaert prefers intervention over laissez-faire, bottom-up over top-down approaches that take account of the identities, interests, and values of those affected thereby, and participatory over authoritarian forms of governance. These interrelated preferences inform his observations, first, that current multi-level governance arrangements privilege free trade and promote polarization and exclusion and, second, that there is a contradiction between the rhetoric of democracy as a key dimension of good governance and increasingly authoritarian/opaque modes of governance at all scales.

In addition, work on multilevel governance and the network polity poses fundamental issues about the extent to which a network polity will remain tightly anchored in territorial terms (as opposed to being necessarily territorially embedded) despite its highly pluralistic functional concerns and its equally variable geometries. Scalar-specific regulations act also as 'articulating filters', that is buffers to ensure the least conflicting articulation among different geographical production/reproduction systems. Paraphrasing Gramsci, who analysed the state apparatus in its inclusive sense as 'political society + civil society' and saw state power as involving 'hegemony armoured by coercion' (1971: 261-3), we regard the state apparatus as based on 'government + governance' and as exercising 'governance in the shadow of hierarchy' (Jessop, 2016: 9).

GOVERNING THE SOCIAL ECONOMY

A key aspect of research on governance is (or should be) how problems are socially and discursively constituted. This is seen in Moulaert's advocacy of the social economy as a solution to economic and political challenges and, more recently, linked initially to the DEMOLOGOS project, in cultural political economy (Sum and Jessop, 2013; Moulaert et al., 2007b; Moulaert and Mehmood, 2010; Novy et al., 2012). The social economy is now seen to involve distinctive economic and social imaginaries that are oriented to substantive goals, guided by normative concerns, and depend for their local success on specific metagovernance practices. Moulaert defines the essential features of the social economy as follows: 'the social economy is that part of the economy – or the complement to the "co-existing" other economy – that:

1. organises economic functions primarily according to principles of democratic co-operation and reciprocity;

2. generates a high level of equality and/or organising redistribution when needed;

3. satisfies basic human needs in a sustainable way' (González et al., 2010: 54, citing Moulaert and Nussbaumer, 2005).

Note that this definition indicates that the social economy is part of a broader economic order. There is also a strong Polanyian element here because Moulaert and his various co-authors stress that the social economy is also a social market economy. In other words, market forces still have a role in providing needs-satisfying goods and services; but, in good metagovernance fashion, this role is subordinated to the requirements that there is voluntary cooperation within and across enterprises, fair compensation for all factors of production in the value-chain, reciprocity in exchange, and a high social and ecological quality in goods and services with minimal negative impact on society and the environment (González et al., 2010: 55; cf. Polanyi, 1944). This requires a continuing process of collibration to get the balance between modes of production, distribution, and governance right, with priority given to the social economy, social cohesion, and solidarity as guiding principles and aspirations. In short, for Moulaert, governance and metagovernance are core substantive themes. ◀

Parc L28 © ndvr

New Orleans © Angeliki Paidakaki

19. BOTTOM-LINKED APPROACH TO SOCIAL INNOVATION GOVERNANCE

◆

Marisol García and Marc Pradel

bottom-linked approach to social innovation recognizes not only the centrality of social initiatives taken by those immediately concerned with specific social needs, but also the need for the support of institutions that enable and sustain such initiatives through sound, regulated and lasting practices (Pradel et al., 2013). This bottom-linked approach can be seen as a step forward in social innovation governance. When social actors engage in creative strategies to deal with social needs, they – as agents of social action – become involved in the forms of *internal* and *external* governance. Actors develop internal rules and mechanisms to ensure the effectiveness of their organization and of the strategies they develop to tackle social needs. In doing so, social actors can be creative and innovative not only in the project(s) they undertake, but also in innovating their internal governance mechanisms. In terms of external mechanisms, actors of social innovation need to reflect on what kind of relationships they prefer with the social and institutional environments in which they are embedded. On one hand, this reflection can lead them to stay at the margins of social and political institutions (for example, away from social services departments or city councils).

Alternatively, these actors can engage in discussions with institutions to put forward their proposals, which may involve innovating in governance beyond their organizations. When this process starts, social and institutional actors often debate for some time how to relate to each other and eventually to establish an understanding for coordination or cooperation. If an agreement for cooperation is finally reached, support from the institution(s) to the specific project(s) may help the innovation to become more sustainable. This stance is highly visible in the majority of studies (see Moulaert et al., 2013b). However, the construction of bottom-linked approach was already envisaged years earlier in the Integrated Area Development approach (see for example, Moulaert, 2000).

A bottom-linked approach can involve allocating budgets, forming teams and providing support through legislations. A further step is the constitution of social rights, although social innovation may not always achieve this. Outcomes can be recognition and support from public authorities for innovative civic practices, or recognition of cultural diversity or empowerment of the socially excluded. Other possible impacts of bottom-linked social innovation on policy-making can lead to changes in: (1) the definition of a policy problem; (2) policy-making processes; and (3) policies and their results. Most of these processes often take place at the local level. However, in order to overcome the "localist trap" actors often engage in calling the attention of citizens and institutions at regional and national levels. This *up-scaling* may involve assuming responsibility for maintaining or improving welfare resources and/or strengthening the implementation of social rights at the national level. This also offers new hope in challenging the current economic and political context of welfare-state retrenchment and liberalization of social services. ◀

© ndvr

SPATIAL
DEVELOPMENT
PLANNING

20. FROM SOCIAL INNOVATION TO SPATIAL DEVELOPMENT ANALYSIS AND PLANNING

◆
───────────

Seppe De Blust and Pieter Van den Broeck

The literature and practice in social innovation seems to have a rather peculiar relationship with planning theory and practice. Earlier research on socio-spatial dynamics, produced by the URPSIC (Moulaert et al., 2003) and DEMOLOGOS (Martinelli et al., 2013) programmes, analysed the logics behind uneven urban restructuring and the possibilities for social action, whereas the subsequent work actively engaged with understanding the possibilities of planning and spatial visioning as practices of social innovation per se (Moulaert and Mehmood, 2018). SPINDUS (Segers et al., 2016) and INDIGO (Van den Broeck et al., forthcoming) offer more recent examples of direct involvement with the definition of the practice of planning and its methodology. Here, the practice of planning focuses on broadening the concept of spatial quality and moving beyond the illusion of a single spatial planning profession. Its methodology is considered as a practical understanding of applying transdisciplinarity in planning practice, and a more in-depth framework on the institutional aspects of planning actions as negotiations, shared visioning or mapping.

This chapter tries to make the importance of social innovation for spatial planning theory and practice more explicit. It defines three major contributions of the social innovation approaches to planning and situates them in the current planning theoretical debate. The first section discusses how social innovation research gives a better insight in the institutional embeddedness of spatial planning. Secondly, the transdisciplinary background of social innovation research is described, as well as its focus on collective problematization as a key concept of social innovative planning approaches. The third section elaborates on the multi-actor approach of social innovation research, and how it reconceptualizes planning as a transformative practice not confined to a professional field. In the conclusive section, main characteristics of spatial development analysis and planning are introduced, as a way of constructing a framework for a socially innovative planning approach.

THE INSTITUTIONAL TURN

Social innovation research must be seen as a deeply institutional endeavour. Defined as both process and action driven, 'socially innovative processes are an outcome not only of collective rationality but of institutionalization processes as well' (Moulaert, 2011: 83). The possibilities of structural change can be conceptualized as part of dialectical processes of collective action and changing institutional configurations.

This embedded perspective on spatial change, grounded in a social science-based understanding and view of reality, broadens the analysis of spatial transformations. The socio-political configuration of planning practice is understood in a relational way and considers cognitive, social, political, economic, ecological and discursive/symbolic readings of planning and institutionalization (Gualini, 2001; Healey, 1999; Moulaert et al., 2016a; Van den Broeck, 2011). Through each of these relations, social innovation literature gives a better understanding of the actual opportunities and limits of transformation and the various levels on which agency can intervene. The position of (collective) agency and the position of planning professionals both structures, and is structured by, institutional change.

The institutional embeddedness of planning results in a conception of socially constructed planning systems, their instruments, actors and rules (Van den Broeck, 2011; Servillo and Van den Broeck, 2012; Van den Broeck et al., 2014). As such, it becomes important to understand these either as the replication of a broader hegemonic socio-cultural and technical imaginary (Servillo, 2017; Moulaert et al., 2007a,b,c), or as supportive to counterhegemonic movements.

This specific interpretation of planning practice often clashes with a more voluntarist reading of the planning professional and his/her freedom to act. Although a social innovation perspective can be easily misinterpreted as a theoretical framework that denies the creative freedom and possibilities of planning in imagining and transforming socio-spatial configurations, social innovation research endorses creativity but does this in a relational way. Closely related to, among others, strategic navigation, as conceived by Hillier (2011), and the 'unique situatedness of particular instances of practice', as described by Healey (2009a: 444), Moulaert offers the possibility of strategic action as an embedded endeavour and focuses on institutional difference and selectivity as a way to define critical room for manoeuvre (Moulaert and Mehmood, 2010; Moulaert, 2005). Those that understand the configuration of power differences and socio-spatial transformation through time, can use this knowledge to define possibilities for social action, when fully aware of the interdependencies between different actors and the multiple ways to intervene (for the bad or for the good and not exclusively limited to a professional field). To fully understand the consequences of this for planning practice, it is important to get deeper into how social innovation literature conceives the dynamics of knowledge and action (Section 3) and how these are closely related to a radical multi-actor perspective (Section 4).

KNOWLEDGE AND ACTION

In one of the few texts where social innovation scholarship directly positions itself vis-a-vis the theory and practice of strategic spatial planning it defines the pitfall of taking 'a too sequential view of the relationships between visioning, action, structure, institutions and discourse' (Moulaert, 2011: 102; Moulaert et al., 2016a). This voluntarist and sequential reading of spatial transformation is positioned in opposition to a dialectical view on collective action and processes of institutionalization that can be critically observed in a number of socially innovative practices. Through a deep analysis of how a certain initiative or strategy reveals and challenges the consequences and opportunities for a more equal spatial transformation, one can understand the path dependency of the resources for (institutional) transformation and actively challenge them (Moulaert et al., 2007b).

This dialectical approach to knowledge and action needs to be understood within a sociology of knowledge approach (Moulaert and Van Dyck, 2013) and a pragmatist and holist epistemology (Moulaert and Mehmood, 2014). Pragmatism, as a shared epistemological framework for many planning scholars (Healey, 2009b), stresses the need to acknowledge the close relationship between transformative action and a deeper understanding of possibilities for socio-spatial transformation, and considers an ethical and methodological connection (process-oriented, feedback relationships, and so on). Holism further engages with this multi-layered approach by fully integrating the diversity of social relations while showing their common features and as such creates space for a more bottom-linked approach to social innovation and planning (among others Miciukiewicz et al., 2012). Holism can thus be seen as a preferred research method to understand social transformation, which closely relates to a specific view on planning as an action-oriented endeavour, with an explicit normative focus and the recognition that systematically produced knowledge (taking into account its political-institutional context) has value in shaping and evaluating interventions in the practical world (Moulaert and Mehmood, 2014: 97).

Different methodological applications of this holist/pragmatist epistemological framework have been further explored in social innovation literature. One of the main approaches that can be found and applied in planning practice is the idea of and radical engagement with collective learning as a process. Following a pragmatist vision of the interaction of research and action in which analysis, planning, design and (social) action go hand in hand, collective learning further specifies this by adding an ethical and empowering perspective (Moulaert et al., 2013c). Collective learning can be seen as a transdisciplinary endeavour where, by looking collectively to causal powers and mechanisms, a diversity of life-worlds and scientific perceptions of problems can be considered (Cassinari and Moulaert, 2014; Moulaert and Nussbaumer, 2005). In order to structurally embed collective learning in planning practice, collective problematization should be seen as a clear methodological tool to guarantee the inclusive and transdisciplinary character of collective learning processes. Collective problematization is a specific configuration where all involved actors try to jointly define the research questions, a meta-theoretical framework and the action research methods used in the socio-spatial research trajectory as part of an encompassing planning process (Novy, 2012). In relation to planning practice, collective problematization bridges the theory-practice gap by providing a clear methodological and ethical answer to the transformative but institutionally embedded position of planning practitioners (Segers et al., 2016). Highly embedded in a social innovation-based epistemology, collective problematization acknowledges the direct involvement of a local needs agenda and the possibility for empowerment by changing existing social relations. In order to fully understand the consequences of this way of working and how it relates to the strategic character of spatial planning and the complex social reality within which planning activities are embedded, it is important to fully understand how this logic is part of a wider multi-actor interpretation of planning.

A RADICAL MULTI-ACTOR APPROACH

It makes no sense to try and understand the implications for planning practice of a dialectical institutionalist approach and collective problematization without considering a radical multi-actor perspective on planning. The understanding to take social innovation analysis and practices as a start to define the room for manoeuvre of a more egalitarian urban development, implies the recognition of a diverse set of spaces of opportunity to positively reconfigure the underlying development dynamics for the better. Social innovation research rejects the exclusive character of strategy, creativity and action and fully integrates the idea of differential agency into its theoretical endeavours. The recognition of the power of socially-innovative practices must first-of-all be understood as the recognition of the mobilization of selective strategies by all (and not only professional planners) (Van den Broeck, 2011).

By approaching planning practice and theory from a social innovation perspective, the concept of governance goes far beyond established discourses and practices of technical/legal regulation and a mere technical-rational use of instruments. The multi-actor approach regarding complexity, scale, time and involved actors expands the space for innovative actors. It engages with the strategic question of planning as a transformative practice (Albrechts, 2011; Healey, 2009a; Friedmann, 2011) that isn't limited to a professional field. In order to seriously engage with a multi-actor perspective on planning, one should understand how transformative practices relate to the emergence of new coalitions, changing dynamics of positioning and their implications on (new) types of instruments that can become embedded within an existing institutional context.

From this perspective a socially innovative type of spatial planning 'opens up potentialities for people-to-come (Hillier, 2007: 232) and is based on the capacity of human beings – as a response to problems, challenges and potentials – to create, improve and reshape their places with the aid of knowledge (scientific as well as local), innovation and transformative practices that work with history and overcome history' (Albrechts, 2011: 20). A multi-actor approach integrates the possible multiplicity of engagement and goes beyond the planning professional ideal of collectively building a shared ontology. From a multi-actor and social innovation perspective on transformation and planning, institutional selectivity, collective action and its embedded necessity of interpersonal solidarity are key to understand the possibilities for action.

SPATIAL DEVELOPMENT ANALYSIS AND PLANNING

Each of these key characteristics of a social innovation-based perspective on planning gives clear ideas on the basis and methods modernizing the theory and practice of strategic spatial planning. The institutionalist ontology, the sociology of knowledge epistemology and the radical multi-actor interpretation of planning, have provided the basis of spatial development analysis and planning, as a new generation of strategic spatial planning approaches. It could be defined as a range of activities that may contribute to local or regional territorial development processes in which the needs and assets of a range of actors, including underprivileged ones, are collectively addressed in ecologically sustainable, socially just and empowering ways. Based on the previous, a number of its interconnected characteristics could be summarized as follows.

- Spatial development planning includes collective socio-spatial action, addressing needs (access to healthy food, healthcare, pure air, clean soil, safety, legibility, vitality, comfort, ...), and mobilizing community assets, to be defined collectively and for all, including non-conventional or underprivileged actants, with special attention for ecological processes and their social mediation (social innovation, environmental justice).

- It is grounded in a social science-based socio-spatial understanding and view of reality, in which planning and action are socio-politically (that is cognitively, socially, politically, economically, discursively, symbolically, and so on) embedded. In other words, planning and action should be practised from the deep awareness of being embedded in an institutional field of dialectically and dynamically connected actors and institutions, which mediate the local expressions of macro-structural factors.

- This field is also socio-ecological and physical, that is nature and materiality are constitutive dimensions of reality, both as agency and as structuring mechanisms, although they must never be addressed in isolation from the ecological, socio-political, socio-economic, and so on dimensions.

- As a fundamental human perspective, space matters. Spatiality, place-making, place-based and spatially oriented collective action, spatial scales and scalar territorial dynamics provide starting-points for integrated socio-spatial understandings of reality and planning activities.

- A focus on action, grounded in a socio-spatial understanding of reality, implies mobilizing (the interaction of) multiple research and action methodologies, in which analysis, planning, (institutional) design and action go hand in hand in a dialectical, iterative and multiple process of gradually building ideas on solutions, projects and new ways of problematizing the existing situation. Planning should be grounded, pragmatist, relational and methodologically holist. As such it is deeply socially constructivist.

- Territorial development implies radical socio-spatial transformation, thus requiring an understanding of time, history, path-dependency and path-creation as well as the temporalities of actants and institutions, how these shift, are reorganized and interact.

- To a large extent, spatial transformation can be a collective learning process, built by solidarity and cooperation seeking actors. However, reality includes diversity and differential agency and structuration, which embeds planning activities in power structures and power relations, and mechanisms of inclusion and exclusion. Capacity building, empowerment, social mobilization and socio-political transformation should be deeply embedded in the planning perspective, as inspiration as well as goal.

- Differential agency and structuration lead to the mobilization of selective strategies by all, including planners. Knowledge of planning and design strategies deployed in the past, their contexts, as well as the positioning of their related actors, is therefore useful.

- Planning theory and practice are embedded in the socio-ecological reality. Integrated approaches, connecting various actants, needs, assets, strategies, institutional frames, spatialities, temporalities, scales, methodologies, and so on, including institutional design, should be favoured.

These features are obviously not clear-cut, raise many questions and need to be operationalized in concrete situations. It is up to the actors involved in these situations to create their own meta-frameworks for problematization, conceptualization and intervention. Consequently, the ethical, socio-political and organizational positioning of those who aim to play a role in setting out trajectories for spatial transformation (planners included) in a situation at hand, is key in the approach to spatial development analysis and planning. As such, whose needs, assets, spatialities, temporalities, projects, plans and so on are included, excluded, privileged, undermined, or involved in suggested compromises, is a question that drives spatial development analysis and planning and that needs permanent monitoring and evaluation.◄

21. RADICAL STRATEGIC PLANNING MEETS SOCIAL INNOVATION

◆

Louis Albrechts

A more radical strategic spatial planning approach (see Albrechts, 2015, 2017 for references) responds to the key challenges in strategic planning. This includes questions on how to embed the fundamental ethical premise of inclusion in strategic planning; and consequently, how to open societal relations for all communities in strategic planning processes and how to put the development of deprived groups and communities at the top of the political agenda. Besides the obvious focus on *physical* space, radical strategic planning has a focus on social space. In social space, socially innovative relations are both catalysts and outcomes of struggle between claims over physical space. In this way, radical strategic spatial planning enters into conflict with political regimes and develops skills that are able to combine a strategic grasp of the contextual dynamics of particular challenges and situations with a real awareness of the particularity of moments of opportunity in which cracks can appear in institutionalized discourses and practices.

Moulaert (2011) explores the common ground between social innovation and strategic transformative practice. He identifies similarities between the transformative agenda as described by Albrechts (2011: 20) and social innovation as collective agency. The narrative of radical strategic planning is a narrative of emancipation (or, radical social innovation, per Moulaert, 2016). By bringing into policy deliberations, the voices and views of those living in an area and experiencing its problems, it fulfils a legitimating function. It legitimates social and political institutions and practices, forms of legislation, ethics, modes of thought and symbols. It grounds this legitimacy not in an original founding act but in a future to be brought about, that is, an idea to realize. This idea (of equity, fairness, social justice) has legitimating value because it is universal. So, apart from legitimacy stemming from a representative mandate, in radical strategic planning legitimacy may come from its performance as a creative and innovative force and its capacity to deliver positive outcomes and gain benefits for all communities. ❮

Schorvoort © Jean Schreurs

22. DESIGN AS SOCIALLY INNOVATIVE CO-PRODUCTION

Ahmed Z. Khan

The notion of *design with community* highlights a take on design as a field of knowledge and practice to engage with as a social scientist in inter- and trans-disciplinary ways. In various design-related research projects already mentioned elsewhere in this book, design as a discipline and designers of many streaks, have been engaged. The shift in understanding *design* as an autonomous discipline to that as an integral part of the cultural process in shaping the built environment is something I owe to my collaboration and experiences with my former colleagues in the Planning & Development Research Unit at KU Leuven, more specifically in the SPINDUS research project (Segers et al., 2016).

Design and imagination. Design is a broad canvas of creativity, imagination, and a projective form of knowledge production and practice; from the design of nature and landscapes – the functioning of ecological systems – to the design of social systems, lifestyles, ways of production and consumption, objects and products, services, and inhabitations of space. Design in an architectural and urban sense, however, has been dominated by the narrow functionalist confines of the modernist doctrine for most of the twentieth century, where the debate 'revolved around issues of how the form of objects could enhance the quality of life' (Margolin and Buchanan, 1995: xi). This has been reversed, thanks to interdisciplinary contributions from the social sciences and humanities. The focus now is on the 'psychological, social and cultural contexts that give meaning and value to products and the discipline of design practice' (ibid: xi). From this reversal, design is seen as a field that engages and enquires into 'the human experience that the built environment evokes' (Sternberg, 2000: 266), and thus engagement with the cultural, socio-economic and environmental conditions affecting our quality of life becomes integral to the design process.

In this sense, design is a way of thinking – a habit of imaginative playing with possibilities in complex and ambiguous situations – and a result oriented attitude that aims to analyze the problems and challenges of specific contexts, explore potentialities and possible transformative solutions, and, in short, a 'principal method used by society to envision how we want to live in the future' (Moore and Karvonen, 2008: 30; cf Khan et al., 2014).

Design imagination and social innovation. Cognizant of these shifts, and adamantly against the role of design or designer as an *autonomous* entity no matter how intellectual, enlightened or benevolent s/he might be, Frank Moulaert contributed towards mainstreaming an understanding of design as a user-oriented, collective creative enterprise that has a societal relevance and meaning, that is design as a socially innovative co-production. But what does a socially innovative design look like? What would it take to make design a socially innovative co-production?

Socially innovative design implies three steps: 1) shared definition and satisfaction of needs of the users/local community; 2) their empowerment in fulfilling those needs; 3) leading to improvement in their social relations unfolding inclusive and sustainable development. The first step requires not only active participation, but also paying attention to power relations, gender issues, and so on. It also implies design as a team effort; designers, ecologists, sociologists, and other specialists working together with the local community. The second step is to view governance relations and institutional arrangements as collectively constructed bottom-linked initiatives that give socio-political empowerment and a sense of ownership to the local community, which is crucial for active and responsible citizenship. The third step is about realizing good community living where improvements and innovations in social relations is a dynamic process for building social capital, local identity and enhancing solidarity.

In short, underlying these methodological apparatuses is the main idea: to make design imagination work together with social innovation for community wellbeing and unfold inclusive, resilient and sustainable development. ◄

© ndvr

23. MOVING IS KNOWING. KNOWING IS MOVING

Jan Schreurs

Walking is not merely bridging or connecting distant locations. Walking is both moving and being on the move, thus *becoming*, both bodily and mentally. Walking is creating. By moving along paths, through nature and cities, knowledge is forged. Not unlike the Situationists' *dérives*, walking is a rich physical and artistic experience.

This I learned walking Schorvoort, a neighbourhood of Turnhout, and a case study in the SPINDUS research project (Segers et al., 2016). SPINDUS, Spatial Innovation, Planning, Design and User Involvement, aimed to — examine social innovation and spatial quality, concepts seemingly belonging to quite separate disciplinary realms. Since we coined the combination *spatial innovation* however, a common agenda was set. Interrelating and mutually reinforcing the multiple dimensions from spatial design, social innovation and strategic planning, and confronting the conceptual results with empirical cases, served an ambitious aim: the development of practical and pedagogical planning and design methodologies to assess, evaluate and implement spatial quality (Segers et al., 2016). The research focused on the potential of *research by design* as an interdisciplinary and transdisciplinary approach to addressing spatial quality.

In Schorvoort we learned from various ways of thinking, practices and concerns by walking multiple lines. While walking, talking, observing, dialoguing with stakeholders within the multiple concrete contexts of the brief for densifying the neighbourhood, we exchanged methods, opinions and values, learning by doing, making and taking time, while simultaneously progressing physically and mentally from locus to locus, without a predefined trajectory. Gradually developing a *shared language*, I understood that a vibrant social imagination is needed to produce spatial qualities — mentally, socially and physically. Another result is the awareness that process is a product too. Not unlike living organisms, participatory processes thrive on fragile metabolisms. Nourishing flows of transformational ideas is one of the crucial skills, to be developed and distributed among researchers and stakeholders alike. ◄

10

SOCIOLOGY OF KNOWLEDGE

24. TRANSFORMATIVE SOCIAL INNOVATION, CRITICAL REALISM AND THE GOOD LIFE FOR ALL

Andreas Novy

When social innovations are seen as practices that improve human conviviality in a broad sense, going beyond a materialist understanding of development and growth, aiming at 'transformation of institutions, overthrowing oppressive *structures with power*, supporting collective agency to address non-satisfied needs, building of empowering social relations from the bottom-up' (Moulaert and Van Dyck, 2013: 466), then theory and practice have to be related, as action and reflection are intertwined. Social innovation research, which contributes to transforming social relations, needs good theories, good methods, but especially a good research design. With respect to apparently problematic situations, like urban degradation, unemployment or intercultural conflicts, it is important to know the *fundamental social relations without which this phenomenon would cease to exist*. This recalls the old saying that 'when it comes to practicality, nothing beats a good theory' (Danermark et al., 2005: 187f) – and a good theory of capitalist modernization is prerequisite for all types of emancipatory agency. Adequate practical knowledge and transformative social change are related and need one another (Moulaert et al., 2013b).

In the project on *Social Cohesion in the City* (SOCIAL POLIS),[1] we proceeded from interdisciplinary teamwork to deliberate efforts of transdisciplinary knowledge generation and diffusion. To *grasp the full picture* of a complex issue like social cohesion requires widening our perspectives, reframing problems and mobilizing different theories and methods. Inter- and transdisciplinary research designs contribute to broadening the actor's horizons by integrating practitioners into the research process, thereby obtaining a deeper understanding of *reality*. Social cohesion, for example, cannot be understood by one discipline, nor can it be dealt with by socioeconomic, political or cultural policies alone. It requires multiple perspectives and approaches to tackle the *problématique* of striking the balance between equality and diversity, belonging to a group and distinguishing oneself as an individuum (Novy et al., 2012). It contextualizes itself in many wicked problems, like increasing inequality and widespread social insecurity, urban degradation in the French *banlieu* or the security issues related to the refugee tragedy. Resulting questions are not easy to answer, and definitively not de-contextualized: which cultural traits should be respected, which prohibited; which elements of inequality foster diversity and which undermine cohesion?

Over the last years, however, social innovation has increasingly been reduced to a universal and homogenized recipe of fostering social entrepreneurship and creating quasi-markets (Jenson, 2015: 101), thereby creating an *enabling welfare state* which uses the creativity and personal commitment of citizens (BEPA, 2010: 7). In current social innovation policies, attention focuses on the space of manoeuvre of deliberate agency, often by social entrepreneurs, to implement *piecemeal changes* in the short run, like improving language skills of migrants or reintegrating long-term unemployed into the labour market. Nobody can object to *doing more with less* in the form of cost- and resource-efficient responses in times of resource depletion. Nor can one oppose incentives for active citizenship. However, these efforts have become increasingly problematic, as a one-sided concern with measurable social impact, offering quick and visible solutions, has impeded attempts to reflect on the deeper causes of the current multiple crises.

To deal with such wicked problems, as phenomena which can be changed by human agency, requires an adequate understanding of the world. Critical realism as an embracing philosophy of science insists that reality exists independently from observation and conceptualization by human beings. But it has to be approached critically and in awareness of the variety of concepts to perceive it, as the real world is different from our descriptions of it, reality from its conceptualization (Danermark et al., 2005: 5). The empirical phenomenon of poverty, observable and measurable, must not be conflated with the reality of a complex phenomenon of deprivation, alienation,

---

1. For further information see http://socialpolis.eu/

humiliation and oppression, as specific experiences and events are inherently limited and theory-laden. Therefore, a realistic understanding of the world should be always critical about the prevalent concepts and theories which might lead to wrong – because simplistic – representations of reality.

In ImPRovE, a European research project on social innovations to combat poverty, we discovered a range of initiatives whose social impact consisted in "humanizing" a problematic situation without tackling the deep structural causes (Oosterlynck et al., forthcoming). An emblematic example is the *Energy for All* programme in Flanders which organized a micro-credit programme for energy poor households. It gave incentives for *moderate* indebtedness and opened a new business model for the financial sector, without problematizing the finance-dominated accumulation regime (Cools and Oosterlynck, 2015). Under certain conditions – for example urgent deprivation – these minimalist innovations might be positive in themselves. But it is the opposite of innovative if they deepen path dependency of a failed socioeconomic model of radicalized competition, dismantling public bodies and social rights and fostering consumerism. It is a mistake to accept neoliberalism as without alternative and take 'modesty for realism' (Unger, 2015: 236), as this has often led to 'dull repetition' (Sum and Jessop, 2013: 133) of a cost-reducing optimization logic without changing the causes of social exclusion and ecological degradation. 'To accept the present political and economic arrangements as the unsurpassable horizon within which the social innovation movement must act, is to reduce the movement to the job of putting a human face on an unreconstructed world' (Unger, 2015: 236). Social innovation should not only be efficient with respect to micro-agency but innovative with respect to the dominant institutional setting. Social innovation is incompatible with all types of *There is no alternative*.

Exploring alternatives requires a specific epistemology 'understood as an interactively unrolled manual on how to connect questions about social change to scientific interrogation' (Moulaert and Van Dyck, 2013: 467). Transdisciplinarity helps to overcome initial, often limited and superficial problem definitions. By organizing time, space and resources, it offers space for dialogue and deliberation to jointly define problems and elaborate a research programme (Cassinari et al., 2011; Novy et al., 2013). But social innovation requires more, based on a deeper understanding of reality. In a complex setting with an unclear definition of problems, like shelter for the homeless or respect for diversity under the exigencies of employability, social innovation should be about redefining problems by reframing and, thereby, creating different *realities*. This is much more than problem-solving. Instead of problems, there are – in a more neutral language – phenomena which have to be understood: unemployment due to productivity increases might be transformed into a blessing with working time reductions; migration might be re-interpreted as a growth potential instead of a security threat; poverty might offer insights into a sustainable way of life. Creating the context in which the potential of a concrete situation becomes actualized goes in a creative way beyond the *actually existing*.

Critical realism and the type of social innovations which has inspired a maximalist understanding have their ideational origins in the progressive mood of the 1960s, perceiving the future as open. As reality covers the existing as well as potentially emerging, critical realism conceptualizes the potential of a problematic situation as part of reality, broadening the scope of *realism* in policy making (Moulaert et al., 2013a; Oosterlynck et al., forthcoming). Thereby, it emphasizes the potential of human agency for path shaping and opens a whole universe of emerging futures. This transformative type of social innovation opposes the currently dominant type of social innovation research and policy as actualist, that is reducing the real to what exists at the moment, thereby excluding potentially progressive futures. This has been the fate of many of the current initiatives promoted under the Barroso Commission. Against actualism, one has to insist on the openness of the world and the pulse of freedom which must not remain trapped in superficial or even ineffective adaptations of existing mechanisms. In the case of *Energy for All*, this would imply empowering poor households to learn to perceive the structural causes of uneven access as well as the political production of indebtedness, in times when public money is available to save bank shareholders but not ordinary citizens. For sure, this would question the existing governance model of many social innovations, which is increasingly based on cooperation with business or philanthropy. And it shows the urgency of a power-sensitive conceptualization of social innovations which explores spaces of manoeuvre for empowering and emancipatory agency in adverse and conflict-ridden situations.

The real is neither limited to empirically observable nor actually existing phenomena, but constitutes the underlying, non-observable mechanisms that can cause events in the world too. Reality – being observable or not – imposes structural constraints and opportunities for the involved agents comparable to natural laws – like gravitation or mortality. These non-observable mechanisms can only be grasped with adequate concepts and models. Without the concept of exchange value one cannot understand the ongoing economization of social services and the subordination of the multi-dimensional use value to a mere by-product of successful economic activities. Ignoring this would underestimate the power of profit-making in market societies with its disastrous consequences on the human fabric, described historically by Polanyi (2001[1944]), and currently by Crouch (2015). Nevertheless, these unobservable mechanisms – like the growth and profit imperative in capitalism – exist and constrain, but also facilitate human agency. Knowing them is decisive for effective transformative agency. Such a focus on structures as both enabling and disabling is a prerequisite for social innovation research. 'Social structures are always the context in which action and social interaction take place, at the same time as social interaction constitutes the environment in which the structures are reproduced or transformed.' (Danermark et al., 2005: 181).

A focus on potentialities is even more important in times of change. Not only structural unemployment, climate turmoil or the ongoing debt crisis of households and public bodies, but also the rise of right-wing extremism, nationalism and European disintegration show the deep and multiple character of our civilizational crisis (Novy, 2014). We seem to be in another Great Transformation, comparable to the transition from an agrarian to an industrial society. Transformative social innovation, therefore, has to dwell – in the words of Polanyi (2001 [1944]: 44) – with the morphogenesis from a caterpillar to a butterfly. Although the same animal, the emergence of a butterfly reflects a profound change in form. In times of transformation, it can be expected that social forms and institutions of living, working and caring will change profoundly. In socioeconomic systems, these changes are always related to power and a respective correlation of social forces. Understanding innovations requires theories that do not superficially reproduce existing social forms, but perceive deep structural inertia. With respect to these structures, agency can be reproductive or transformative (Danermark et al., 2005). Supporting the empowering and transformative dimensions of social innovation has always been stressed by the research tradition embodied in the contributions in this book, thereby contributing to the 'reflexive transformation of structure by agency' (Sum and Jessop, 2013: 49).

Social innovations, conceptualized within Critical Realism, contribute to overcome the destructive and self-restraining dualism of incremental and radical change which has so far impeded the effective mobilization of the marvellous constructive energies of small-scale bottom-up initiatives. But neighbourhoods alone cannot save the city (Moulaert et al., 2010). The challenge consists in avoiding that micro-social innovations fall into the localist trap or even foster institutional lock-ins.

Conscious collective transformation means learning to change social forms, institutions, discourses. '[T]he point of all science, indeed all learning and reflection, is to change and develop our understandings and reduce illusion. ... Learning, as the reduction of illusion and ignorance, can help to free us from domination by hitherto unacknowledged constraints, dogmas and falsehoods.' (Sayer, 1992: 252). New and better theories have to be elaborated and adequate concepts have to be generated to grasp important aspects of the context which shapes bottom-up initiatives. To give one example: to reduce the political to state agency depoliticizes all efforts to democratize the social-ecological transformation, be it the energy transition or new forms of mobility. Transformative social innovations broaden freedom, not in the sense of increasing freedom of choice, but as the capacity to shape one's own way of living. It is based on democracy as a form of life, on participation and public involvement in the community and public affairs. It is the *Gemeinwesen*, be it the polity, local community or the broader commonwealth which creates the infrastructure and institutions which hinder and enable agency: bicycle lanes or motorways, local markets or shopping malls, free access to the internet or big-data controlled social media. Only such a broad concept of the political allows the transition movement and other alternative social movements to make sense of their piecemeal initiatives as steps towards systemic transformations. At the same time, the state, its legislative and tax-collecting power, remains a nodal point in framing the context for such a broadened political agency.

During post-war social-democratic welfare capitalism, social policy has been efficient as state-centred, top-down and homogenizing. As this type of social engineering has been dismantled and discredited for good – and not so good – reasons, there is an urgent need for a new progressive method for transformation that links visionary thinking of *another world* with actual reality and pragmatic, often grassroots agency in a way that is practically adequate. To deal constructively with such long-term transformations requires *maximalist* social innovations, 'piecemeal and gradual in method but nevertheless radical in ambition' (Unger, 2015: 239). The concrete utopia of the good life for all aims at the free flourishing of each as the condition for the free flourishing of all. This opens up a comprehensive utopian horizon beyond capitalism, while providing guidance for reform strategies effective in the short-term, within the existing order that has solidified neoliberalism over decades in Europe. 'As a guiding principle, it avoids the dualism of reform and revolution, of small steps within the existing order and major advances of radical change' (Novy, 2014: 5). This is an emancipatory method that explores the potential of bottom-linked transformative agency. Structure-aware agency valorizes effective pragmatic first steps in the direction of radical change foreshadowing a different future. 'Such steps are moves in the penumbra of the *adjacent possible* surrounding every state of affairs: the *theres* to which we can get from here, from where we are now, with the materials at hand … Only because the piecemeal can be the structural can the social innovation movement do its work.' (Unger, 2015: 242). But piecemeal must not be confounded with bottom-up or project-centred, favourites of current utopian thinking. The type of social innovation most in need today is the discovery of a new form of political agency which does not limit the political to the state. Political – in its original meaning – is about shaping the common wealth, its laws and institutions. Politicizing – in small steps – transformative agency as a collective undertaking of overcoming the deep logics of consumerism, endless competition and the growth imperative is the major challenge of our times. Transformative social innovation has much to contribute. ◄

25. TRANSDISCIPLINARY PROBLEMATIZATION

Konrad Miciukiewicz

Transdisciplinary problematization is a process of critical consideration of problems and challenges to be addressed through problem-focused research and transformative practice. It is the first step of a transdisciplinary endeavour whereby a transdisciplinary team (for example researchers, citizen scientists, policy-makers, entrepreneurs, activists, community leaders and interested individuals) engage in collective deliberation over the nature of problem(s) they wish to solve, the symbolic and material resources they will mobilize, and the ways in which they will relate to each another throughout an interactive research and practice project. As part of the transdisciplinary problematization, actors discuss issues to address, aims to pursue, and questions to research; they talk about different academic, professional and experiential knowledges they will use and develop to understand the problem(s), and about tools they might put to work in the transformative practice; they also think about roles that would emerge throughout the process, who would play these roles, and how the different actors and actions would interact.

Transdisciplinary problematization has three methodological aspects. First, it is a dynamic and critical process whereby actors with different desires and aspirations for, and roles in, public life can challenge the dominant discourses, evoke new viewpoints, and co-create novel approaches and practices (Novy et al., 2012). Second, this problematization involves thinking carefully about understandings of broader societal structures, the ways these structures emerge and are reproduced, and forms of shared ethics (Sen, 2001) and ethical agency that can bring about social change (Moulaert, 1987; Moulaert and Jessop, 2013). Third, the transdisciplinary problematization leads to a collective negotiation of a *meta-theoretical framework* that draws upon concepts from various disciplines, addresses multi-scalar and historical socio-spatial relations, and is capable of "hosting" contradictory relationships of socio-political reproduction (Miciukiewicz et al., 2012; Moulaert et al., 2013c). ◄

26. CURATING THE META-FRAMEWORK

Michael Kaethler

The famous saying *writing about music is like dancing about architecture* emphasizes the futility of formal language when engaging with the arts. Theodor Adorno (1997), in his work *Aesthetic Theory* also makes a similar distinction, between mimetic reasoning, where one assimilates himself (or, herself) towards an object of study, and instrumental reasoning, which demands a stepping back, an analysis, and deconstruction. The former denies the latter and vice-versa; each is a unique mode of reasoning. How then, given the distinction between art and analysis, can one develop a framework with practitioners which explores creative practices while also reflecting upon them through theoretical models?

Taking inspiration from the SPINDUS meta-framework (Moulaert et al., 2013c), I began delving into theories around tacit and embodied knowledge along with relational epistemologies and ontologies. I was intrigued by SPINDUS' relational ontology and pragmatic frame of reference, which held together different types of knowledge without smothering it with excessively prescriptive theory and analysis (Khan et al., 2013). With SPINDUS as a point of departure, I developed a mode of reflection that employed spatiality and performance as ways in which users could think *with* and ultimately *about* creative practices. It was a type of meta-framework transformed into an exhibition, whereby relationalities became spatialized and knowledge was a performed act of exploring proximities, juxtapositions and rapprochements. It enabled a plumbing of practice and theory without relying heavily on formal language. In this, it provided an accessible medium for exploring practices, without a reliance on language, but instead on relationality. The SPINDUS meta-framework inspired seven exhibitions across five countries, elaborated as part of the TRADERS research project (http://tr-aders.eu/). ◀

27. COLLECTIVE LEARNING

Annette Kuhk

Designing planning processes as collective learning trajectories holds a promise to recognize the complexity and wickedness of contemporary planning issues (cf. Bertolini, 2010). Collective learning has a relatively open trade-off between *lecturers* and *learners*, *organizers* and *participants*, *experts* and *laypersons*, resulting in a recalibration of these *distinct* roles, and possibly also the questioning of power relations (cf. Healey, 1997). This openness is typical for inter- and transdisciplinary approaches (Moulaert et al., 2013b) and raises hope for broader social engagement with complexity, yet also for *better* outcomes, which are more just, socially innovative or sustainable. *Actively co-constructing researchers* are challenged to reflect on the subject of concern (for example just, socially innovative and sustainable solutions to complex planning problems) as well as on the methodological issues of non-linear collective learning trajectories in which they are engaged. Understanding planning as a collective learning process equally acknowledges socio-psychological factors, which help explain the flow of processes. Collective learning helps in dealing with challenging and changing combinations of personalities, and a shifting blend of competences. Also, individuals may take different roles in a process which coalesce, conflict or contrast. For instance, they can be key individuals, specialized advisors, or holistic thinkers, whereas others defend the needs of influential organizations and specific communities or remain rather silent observers of the flow (Brown and Lambert, 2013; Kuhk et al., 2016).

The Territorial Development Projects for the North of Brussels and the central area of Limburg, or the prospective study of the *Metropolitan Coastal Landscape 2100* are recent examples of collective learning trajectories for complex planning issues in Flanders (Kuhk et al., 2016). These complex projects require a certain level of endurance to negotiate the terms, to identify and to engage stakeholders, to keep them committed, to develop appropriate methodologies, to collectively construct knowledge and to gradually build a shared discourse. Such long-spun trajectories allow us to alternate between future-oriented explorations and a focus on implementation. Ideally, collective learning approaches to planning are open enough to create an environment for *guided imagery* as a basis to articulate concrete experiments (Kuhk et al., 2015). As such, collective learning is a non-linear process, similar to the *open innovation* in user-centred living lab approaches (cf. Chesbrough, 2006), which are increasingly applied for urban and other area-based issues. ◄

11

THE SOCIALLY
ENGAGED UNIVERSITY

28. SOCIAL INNOVATION AND UNIVERSITIES: THE CHALLENGE OF SOCIAL TRANSFORMATION

Juan-Luis Klein

In this text, we address the topic of social innovation regarding higher education centres[1]. We seek to identify to what extent universities and other institutions of higher education can contribute to making social innovation the centrepiece of a development model that is more democratic and fairer than the one that currently prevails at the global scale. We are aware of the role that universities can and do play in terms of technological innovation. Concepts such as national and regional systems of innovation, innovative milieus, industrial districts, local production systems or competitive clusters have been presented as combinations of businesses, higher education centres, political bodies and civil society organizations that promote innovative production and competitiveness. However, what we are addressing is different. We are targeting social innovation as such — a concept that does not, at least not directly, focus on productivity or competitiveness but on the collective wellbeing.

From our perspective, social innovation constitutes a social phenomenon that is distinct from productive or technological innovation. The *Centre de recherche sur les innovations sociales* (CRISES) defines social innovation as new social, organizational and institutional arrangements or new products or services with an explicit social mandate that result, voluntarily or not, from an action initiated by an individual or a group of individuals to respond to an aspiration, meet a need, bring a solution to a problem or benefit from an opportunity of action to change social relations, transform a framework of action, or propose new cultural orientations (www.crises.ca).

Social innovations may be incremental or radical; however, what is essential is to view them as milestones in processes in which alternatives are explored at the local level, such as in organizations and communities. Once disseminated, these innovations can contribute to social transformations at larger scales. It should be noted, moreover, that universities and other institutions of higher education are part of the institutional

---

1. This paper is a summarized version of a chapter published in GUNi (2017) *Higher Education in the World 6; Towards a Socially Responsible University: Balancing the Global with the Local.* Barcelona, GUNi, pp: 165-178 (Available at http://www.guninetwork.org/report/higher-education-world-6).

frameworks that often constrain these types of transformations, even if their professors and students take part in the experiences that stimulate them. In addition, these institutions help to produce and replicate those frameworks, namely through the promotion of values and knowledge that shape the society as a whole and that standardize citizens' actions and their ability to analyze problems and aspirations. We believe this to be the level where we must situate the analysis of the role of universities in social innovation, the level of their relationship to knowledge and to the cognitive dimension of the institutional framework.

SOCIAL INNOVATION AS AN ANSWER TO CRISIS

Thinking on social innovation has taken time to develop. However, since the beginning of the twenty-first century, the importance accorded to this topic by major international institutions as well as by researchers and social actors has not ceased to grow (Moulaert et al., 2013b; Klein et al., 2014; Lévesque et al., 2014; Nicholls et al., 2015). This attention stems from the fact that the main technological innovations realized in the productive sectors related to the so-called new economy have led to social transformations that, rather than offering solutions to the problems facing society, have generated or intensified them. Despite a significant increase in wealth at the global scale, the innovations and transformations of recent decades were not driven by a focus on social progress. The flexibility and mobility of capital, the segmentation of the labour market and fierce competition have given rise to new forms of precariousness. All these characteristics of the current development model have been intensified by neoliberal-inspired management reforms of public policy, adopted by governments and applied across all entities of societal governance, including the universities. The 2008 crisis revealed the aberrations of this development model. The fact that the crisis extended over to the social, political and geopolitical realms reveals that the methods available for dealing with such crises are no longer efficient and function at best as palliatives for the major problems of our time, but not as solutions. A paradigm shift is therefore required (Fontan, 2011; Santos, 2011; Unger, 2015).

For most social scientists dealing with this topic, social innovations are new responses to social needs that are not, or are only partially, resolved by the institutional and organizational system in place (Klein and Harrison, 2006; Mulgan et al., 2007; Murray et al., 2010). Most of them acknowledge the important role played by civil society and the social economy in the formation and implementation of social innovations. However, as regards the scope and significance of social innovations, two main perspectives can be identified (Klein et al., 2014). The first one focuses on humanitarian actions for solving specific problems of precarious and vulnerable social groups. By contrast, a more ambitious vision situates social innovations in a broader context where they constitute experiments that can lead to systems of innovation at multiple scales interacting with each other and transforming the society (Moulaert, 2009; Moulaert and Nussbaumer, 2014). From this second vision, which we favour in this text, social innovations may be seen as milestones of a *social movement* (Unger, 2015). According to this vision, social innovation is part of a web of actors rooted in different spheres (public, social and private) and of a social movement that defines itself in terms of inclusive development and that calls for the social embeddedness of the economy (Bouchard, 2013). This vision aims for the improvement of human condition and the quality of life of citizens across all dimensions of human life (cultural, social, environmental, economic, and so on) (Moulaert et al., 2013a), as discussed by Santos (2011) with reference to the *buen vivir* social model conceived in South America.

The transformative potential of social innovations is conditioned by the institutional context in which they operate. This context appears as a set of variables that engage with the internal and external relationships of the actors and which poses constraints, yet that can also promote transformation when the actors create new codes and rules and establish new institutional paths. This latter process, however, is not one-sided. As innovations are diffused and adopted, they evolve through conflicting social relationships and compromises, adapting according to the needs, aspirations and power relations. Reflecting on social innovations as a factor of social change thus requires that we examine the relationships which actors who carry out such social innovations maintain not only with other social actors, who may be allies or opponents, but also with the institutional contexts in which they operate. Essentially, these actors confront and question the institutional framework while at the same time contributing to building and reproducing it.

We point out that the institutional framework is not a homogeneous structure. Instead, it is a set of systems and subsystems of institutions that do not always converge and that have a differentiated permeability to the diffusion and adoption of innovative practices (Unger, 2015). Institutions reflect the social hierarchies, the inequalities between social groups, and the relationships between public authorities and social actors. Educational institutions, and especially those of higher education, are not separated from this institutional context, quite the contrary. It is thus at this level that we must rethink the role of universities so that they may become agents of social transformation through social innovation.

UNIVERSITIES AND THE TRANSFORMATIVE EFFECT OF SOCIAL INNOVATION

Universities are part of a normative and cognitive framework. They validate institutional governance by establishing the values and myths that render them legitimate. Also, the analysis of the cognitive dimension of the institutional framework is crucial to the endeavour of rethinking the social integration of universities with regard to social innovation.

Indeed, universities play a role in the formation and support of social innovations that improve the lives of citizens. However, the analysis of this role must consider the two visions of social innovation presented above. Academic centres and research teams with ties to social actors initiate and participate in localised projects that mobilize knowledge and skills for the benefit of the public good, which promotes social experimentation. They adapt their training programs so as to enable students (and in some cases, through university outreach modules involving citizens) to develop the skills needed for revitalizing devitalized or peripheral areas (Surikova et al., 2015; Nichols et al., 2013; Elliott, 2013). University units are also involved in the transfer of useful knowledge for social experimentation. However, remaining limited to this type of intervention, although undoubtedly important, especially for the communities involved, would move us away from our objective of highlighting the role that universities should play in broader innovation processes aiming for the configuration of new social alternatives.

Benneworth and Cunha (2015) have shown the paradox in which universities find themselves in the context of the crisis of the dominant economic model. On the one hand, their social mission should induce them to actively pursue the development of alternative and solidarity-based approaches and strategies for overcoming the failure of the dominant economic paradigm; yet on the other hand, universities often adhere to strategies defined by global and national bodies that promote elitism and competition, somehow making them promoters of this model. The model of the *world class university*, which most universities have come to acknowledge as the main standard, imposes on them not only teaching and research methods but also values. As a consequence of this model, these universities adopt utilitarian strategies that hinder them from generating and disseminating social innovations likely to change society (Elliott, 2013).

Universities moreover have a major role in establishing the epistemological framework within which development activities are embedded. Through the knowledge produced in them, or through them, they contribute to the definition of what is *real* and *right* and to establishing the legitimacy of actions and actors. As Unger noted (2015: 250), this framework could turn out to inhibit the emergence of alternatives. In other words, institutions of higher learning, to which the institutional framework confers the role of legitimate knowledge producers, tend to impose approaches and methods that consider change only within the limits of the existing institutional framework and that do not call the established order into question.

However, the analysis of the place of universities within the institutional framework also allows us to see the role these could assume in the establishment of an ecosystem of innovation. Thus, by participating in the experimentation with solutions to concrete problems encountered in specific conditions, universities would also be contributing to the vast effort of transforming the conditions that cause these problems and that inhibit the capacity of citizens to change their world, provided, as underlined by Elliott (2013), they make social innovation a strategic priority.

The view that we defend here is that universities, through their bodies of research and knowledge production, can contribute to build a cognitive framework that enables

A KNOWLEDGE-SHARING EXPERIMENT IN QUEBEC

Entitled *Ateliers de savoirs partagés* (knowledge-sharing workshops), this experiment consisted of a project conducted by a team of academic researchers from CRISES and a group of social actors from the Municipality of Saint-Camille in the province of Quebec. Prior to this project, this rural Quebec municipality had already launched numerous social actions aiming primarily at the protection of infrastructure assets whose existence was endangered by social, economic and demographic transformations. The experiment of the CRISES team and the leaders of the municipality consisted of analyzing these efforts. The knowledge co-built by the researchers and stakeholders in the municipality reinforced the community's capacity for action and its power to participate in decisions about its future.

The Mayor of Saint-Camille, some local citizens and some researchers in one of the activities held during the knowledge-sharing workshops © Juan-Luis Klein

alternatives that already exist, but that are either ignored or discredited, to be recognized (Santos, 2011). This implies a paradigm shift in that it allows unofficial knowledge - knowledge of a different cognitive order, co-constructed from diverse knowledge, both academic and practical, and generated, among others, by the stakeholders and actors of social innovation - to see the day. This is the meaning we give to the co-construction of knowledge.

The co-construction of knowledge corresponds to an epistemological vision that considers the relationships between universities and the political, social and economic agents in societies, including civil society representatives, and that challenges the cognitive framework institutionalized by academia and professionals. The co-construction of knowledge calls for the development of reflexivity (Jessop et al., 2013), which constitutes a collective capacity needed for conceiving of new development paths. In particular, it concerns the ability of researchers and actors to imagine new institutional frameworks for social transformation (Fontan, 2011).

CONCLUSION

The current ecological and welfare state crisis is in fact an opportunity to launch a wave of innovations and to co-construct new knowledge between researchers and practice settings to change the existing order (Lévesque, 2014). These innovations should encourage citizens to work towards identifying and fulfilling new aspirations, including the fight against poverty and exclusion, the respect for the environment, the recognition of experiential knowledge, and participation, as asserted by Klein (2015). The current crisis poses a new challenge, since the new wave of innovations will have to be embedded in a new context and ask for a new conception of innovation (Moulaert and Nussbaumer, 2014: 101). The new forms of growth under capitalism generate new divides. While the remaining options appear to dwindle, they do exist and need to be re-examined. The challenge facing the actors, including academic actors, is that of building a cognitive framework, or knowledge framework, that renders these options visible and viable.

Social innovation is primarily based on a collective solidarity-oriented learning process (Moulaert and Nussbaumer, 2014: 105). It thus appears as a necessary ingredient of an alternative development strategy for generating new values. The constant reference to social innovation demonstrates that it is not merely a fad but a prominent feature of an emerging model. However, social innovations will not lead to a new development model unless they are rooted in a unifying perspective that, as advocated by Moulaert and Nussbaumer (2014), gradually shifts its focus away from the resolution of specific local problems and towards a more holistic, comprehensive transformation. The university can contribute to such a shift by producing new knowledge through social experimentation and by disseminating it. The challenge is to produce and disseminate knowledge that is relevant not only for understanding this innovation and transformation process but also for initiating and guiding it. ◄

29. PARTNERING WITH A YOUNG PLANNING SCHOOL: NURTURING A SOCIALLY INNOVATIVE PLANNING APPROACH IN THE UNIVERSITY OF AVEIRO, PORTUGAL

Artur Da Rosa Pires and Carlos Rodrigues

The Planning School at the University of Aveiro, Portugal, has played a pioneering role in the planning education landscape in Portugal, namely through the creation, in 1983, of the first undergraduate degree in Urban and Regional Planning in the Portuguese university system. Furthermore, the structure of the degree was largely modelled on the Anglo-Saxon approach to planning, based on the *applied social sciences* tradition rather than following the *urbanism* tradition largely dominant in Portugal at the time. The University of Aveiro was itself a quite young University, created just a decade before, in 1973. Not surprisingly, this initiative gave rise to some criticisms among the established academy, voiced clearly in a two-day seminar held in Coimbra, in June 1991, about *Teaching and Research in Urban and Regional Planning in Portuguese Universities*, where many well-known professors and urbanists (mainly architects and civil engineers) voiced their scepticism about both undergraduate and social sciences based education in spatial planning, even going so far as to discredit them.

It must be stressed that the University of Aveiro, clearly (and successfully), assumed its innovative stance in many areas and it is now recognized as one of the most dynamic and innovative Universities in Portugal. At the time, it provided solid institutional support to the Planning School, but the relative degree of national "academic isolation" in the planning field meant that international networks were of crucial relevance in the development of the young Planning School. In 1992, the Aveiro Planning School was accepted as a member of the European Spatial Development Planning (ESDP) network, created in the late 1980s by Louis Albrechts, with Frank Moulaert, Flavia Martinelli, Kevin Morgan, Maurizio Garano and Arantxa Rodriguez, among its founder scholars. Becoming a member of this network proved to be a foundation stone of the Aveiro Planning School. It not only provided crucial access to international fora that helped to counterbalance the then relative isolation in national terms, but also had a definitive impact in shaping the epistemological and intellectual identity of the Aveiro Planning School. Indeed, many of the Portuguese students took part in the multiplicity of network activities, and some of them are now quite influential members of the staff while others are active with the alumni network. However, and above all, the influence of the joint research projects that were developed in the context of the multiplier effects of the network proved to be decisive in the very foundations upon which the planning approach of the young school was nurtured, flourished and was eventually consolidated.

We were all influenced by the sensitivity to society that is a distinguishing mark of the work carried out by ESDP scholars. Especially, the pervasive concern with the link between local development and human development is now an unquestioned reference point for our School. The action research ethos – working in partnership with the community in the context of transdisciplinary cooperation, with the aim of contributing to social justice and the capacity of the community to better position itself to effectively address and overcome development challenges, particularly social exclusion – all these traits are present in the ethical, theoretical and methodological foundational principles of the Aveiro Planning School. The renowned concept of Integrated Area Development (Moulaert, 2000), developed in a research project within the EU Poverty III programme, to which the Aveiro School contributed, was a landmark in forming these foundational principles. It also helped to strengthen the need to keep a critical perspective on emerging and potentially trendy or dominant concepts, like some branches of the innovation systems literature that openly neglected the relevance of social innovation as a practice and a process in local development.

Paradoxically, the activities of the ESDP network not only provided a privileged access to international fora and contributed to the densification (and internationalization) of the structuring approach to planning in the Aveiro Planning School but also helped to forge links with the surrounding community in the region. A crucial instrument for this purpose was the Erasmus Intensive Programme (IP), (see Da Rosa Pires, Rodrigues and Cameron in this book), which took place in Portugal for three years in a row, from 2000 to 2002. All three iterations generated a high level of enthusiasm among Portuguese students who provided *host* support to the non-Portuguese speaking students and helped to address four main topics: *Planning Policy, Urban Design and Disadvantaged Groups; Major Urban Projects* and *Urban Change: the Polis Project in Aveiro; The Planning and Management of Major Events in Medium-Sized Cities: the Euro 2004 in Aveiro;* and *Planning in Environmentally Sensitive Areas: Spatial Planning in the Ria de Aveiro Region.* In all cases, students had the opportunity to listen to the regional community and experts and be responsive to their needs and expectations, while also framing those responses in a broader theoretical framework and tailoring them to context-specific circumstances. In this way, this approach to planning that matured over time also actively contributed to root the Aveiro Planning School in its region while providing a far-reaching, global perspective on the nature and purpose of planning. ◀

30. NEWCASTLE:
MAKING THE LOCAL INTERNATIONAL

Stuart Cameron

The transformative influence of social innovation education and research in Newcastle can be traced through two related strands: the way the approach provided new theoretical and conceptual frameworks, and its role in strengthening the international, especially the European, perspective of the School.

The Global Urban Research Unit of the School of Architecture, Planning and Landscape at Newcastle University boasts a strong international reputation in planning theory, with influential thinkers such as Patsy Healey and Steve Graham, joined by Frank Moulaert and Jean Hillier. Frank Moulaert's key theoretical contributions – his individual perspectives on institutionalism and social innovation, for example – are discussed elsewhere in detail, but beyond the power of the ideas themselves was his enthusiasm to debate and challenge ideas within the School and beyond.

As an academic network that combined research with policy and practice through pedagogy, the ESDP network ensured that students from graduate to doctoral levels benefitted from the theoretical tools. Most of all this was seen in the Postgraduate Certificate Programme in Spatial Development Planning. Newcastle has a large doctoral programme in Planning with students from around the world and for many of these students the experience of this programme – the combination of the ideas to which they were exposed and the interaction with students and scholars from around Europe – had a transformative effect on their work.

Three of the ESDP network's major EU Research Framework Programme projects – DEMOLOGOS, KATARSIS and SOCIAL POLIS – were coordinated at Newcastle. Several academics were involved in these projects and together they provided a major impetus to the internationalisation of research in the School, both in its overall output and in the work and academic focus of individuals. I was myself involved in all three of these projects and can perhaps best illustrate the impact in Newcastle through my own research work in and about Newcastle and its region.

Two hundred years ago the city of Newcastle upon Tyne, with its surrounding region, the Northeast of England, was one of the key centres of the Industrial Revolution in the UK, and in the twentieth century it then became one of the first regions to experience de-industrialisation, bringing with it decades of policies responding to the economic,

Byker © Stuart Cameron

Demolition © Stuart Cameron

The famous Byker social housing development in Newcastle is now owned and managed by a tenant organisation, an example of social innovation used in the Katarsis project

social and physical problems this created. The city and region have, therefore, provided a wonderful laboratory for the study of social innovation in urban and regional development and regeneration, and with colleagues at Newcastle University I have taken full advantage of this in my research.

A local case study set in Newcastle and the Northeast of England (Cameron and González, 2013) formed part of the output of DEMOLOGOS, one of eight case studies drawn from across the globe. However, the key purpose of the DEMOLOGOS project was the review and synthesis of urban and regional development theories to provide frameworks for analysis of empirical material. In the case of the Northeast case study, for example, material from a number of local empirical studies was re-framed to explore the scalar and temporal relationships of neoliberal approaches to regeneration. These were traced from the neighbourhood through to national and European levels, for example linking social mix strategies at neighbourhood and urban levels to endogenous growth approaches to regional development. A series of key questions gave the impetus for each research team to apply the DEMOLOGOS synthesis to frame their case studies, not in a wholly standardised way but appropriate to the local case.

KATARSIS and SOCIAL POLIS both used social innovation as a core theme (Moulaert et al., 2013b). My own particular contribution focused not on the Northeast of England but on comparative study of housing and neighbourhood issues and policies across Europe. Conceptualizations of the processes of social innovation – helped by the engagement of practitioners as well as academics in these programmes – provided new insight, for example on the potential of large third-sector social housing agencies as vehicles of social innovation (Martinelli et al., 2013).

Both of these areas of work have provided a rich source of material for my subsequent contributions to teaching, especially my inputs to the programme in Spatial Development Planning. This is very much part of the integration of research and teaching which has been central in the long history of the socio-scientific project of the ESDP Network. ◄

12

LEARNING THROUGH CRITICAL THINKING

31. THE ERASMUS INTENSIVE PROGRAMME: ENJOYING THE CHALLENGES OF TEACHING AND LEARNING IN/FROM A MULTICULTURAL APPROACH TO PLANNING

◆

Artur Da Rosa Pires, Carlos Rodrigues and
Stuart Cameron

The Erasmus Intensive Programme (IP) was an extraordinary multicultural and multi-institutional teaching and learning experience that was run almost every year for a 25 years long period (between 1990 and 2014). The amazing longevity of the IP certainly reflects the longstanding dedication and enthusiasm of all those involved in its organization, but it also signals the huge institutional impact it achieved over all those years.

The Intensive Programme (IP) was a two week-long, structured graduate course funded by the Erasmus Lifelong Learning Programme. The objective was to bring together postgraduate students and teachers from the planning, urban sociology, architecture, urban and regional economics, and geography programmes of the ESDP network in one of the partner university cities. It provided a unique opportunity to mix students from different European universities and disciplinary backgrounds to discuss contemporary planning issues within the framework of applied project work. In every iteration and in the context of each chosen locality, specific spatial planning and development issues selected in advance were theoretically addressed in plenary lectures by both staff members and local experts, and a coherent methodological framework was then provided in order to support the work of students. These were supposed to develop a context-bound critical understanding of those issues and to draft theoretically informed but context-specific planning proposals, development strategies or focused projects. Such task was a major achievement in the context of multicultural and multidisciplinary group work carried out in a very short period. The activities of the IP typically included introductory plenary lectures, site visits, meetings with local stakeholders and quite intensive group work, organized in topic teams as well as report writing and plenary presentations and evaluation. Topic teams were formed according to the preferences of students but also ensuring multidisciplinarity and multinationality.

The Erasmus IP was immediately a major success among students, despite the undeniable heavy workload and the daunting challenges of harmonizing such diverse disciplinary and cultural backgrounds. As one of the participants said, the IP 'left an indelible imprint on students, belonging to different social and cultural backgrounds, engaged in higher (intensive) learning, managing tensions and misunderstandings, working around the clock, but enjoying the social learning dynamics of an ambience of friendship, creativity, and camaraderie'. Indeed, an imprint to which participant teachers were not immune...

This experience and its academic foundations travelled very well between different countries, from Belgium (Leuven), where it all started, to Portugal (Aveiro), Spain (Bilbao), Italy (Milan), Greece (Athens) and Turkey (Ankara) but also to France (Lille), the UK (Cardiff) and Sweden (Stockholm). The timing and location of IPs sometimes created specific learning opportunities. The two IP events in Athens allowed a 'before and after' study of the impact of the Olympic Games on the city and the significance of mega-events in regeneration, while Milan Bicocca – a university created on the site of the old Pirelli tyre factory – itself provided the opportunity to consider issues of de-industrialization, the shift to a knowledge economy and the role of technopoles.

The huge success of this initiative cannot be dissociated from an apparently soft but indeed quite demanding scientific management of the whole programme. Since the very beginning a lot of attention was given to the critical views of students and to a rigorous self-assessment, by the staff, of the adequacy of the academic support and guidance provided. The gradual consolidation of the structure of the IP also benefited from a parallel but (still) ongoing discussion among staff members about the nature and purpose of spatial planning and the intellectual, ethical and prosaic challenges of planning practice. Some of the discussions took place in preparatory meetings. In the memory of most of the current members of the network is a stimulating and quite intense debate in Ferrara, Italy, in January 2002 that was held as a 'platform for discussion about planning and planning education after September 2001'. The network activities also proved quite fruitful in launching research projects and preparing (often successful) joint applications. As a result, the research interests and knowledge produced in the framework of those projects also helped to shape and decisively influenced the approach in specific IP editions. For instance, the extensive and widely celebrated work developed on (urban) social innovation and social justice was a major foundation for the IP editions in Ankara and Stockholm.

Since the very beginning, however, major consideration was given to the actual operationalization of the underlying principles of the IP. For instance, two major issues came quickly into the agenda of discussions: how to accommodate the requirements of multicultural group work, namely through a cross-cultural approach to planning issues, methods and techniques in the context of applied project work; and how to develop the IP so that the results could have, at least, a sense of responsiveness and usefulness to the local institutions which collaborate in the IP. In other words, the academic challenges of the IP were permanently addressed and rigorously considered throughout the IP history.

There were also quite enjoyable and hilarious moments. Flavia Martinelli (2008) wrote a wonderful short text about 'Papa ERASMUS'[1] that is also a testimony to the multifaceted role of Louis Albrechts as the first coordinator of the network. When Louis left his role, the network tried to alternate the coordination among members, but soon one of our most celebrated and warmly appreciated colleagues created the right conditions for the whole set of administrative forms to be (involuntarily) sent to ... the garbage bin! – that was the moment when the network decided unanimously that Frank Moulaert was the one and only prepared to adequately perform such role at the forefront. ◄

_____

1. http://esdp-network.net/intensive-programme-general-information#

32. 'EQUATERRITORIA': A CATALYST OFFSPRING FROM THE EUROPEAN MODULE IN SPATIAL DEVELOPMENT PLANNING (EMSDP)

◆

Chiara Tornaghi, Barbara Van Dyck, Brenda Galvan-Lopez, Daniela Coimbra de Souza, Giancarlo Cotella, Giota Karametou and Pieter Van den Broeck

❛ Either you take yourself seriously and find a theoretical framework that is coherent with your ideals, or you continue what you are doing, but then I recommend you to find yourself a very good shrink. ❜

Frank Moulaert
during one of the EMSDP lectures, somewhere in March 2006

The three months of intensive teaching in the postgraduate programme *European Module in Spatial Development Planning (EMSDP)* have been a life-changing experience for many people like us. We attended the EMSDP in 2006, and life has not been the same since! In this short piece, we hope to give a flavour of how the programme worked, what it generated in our experience and what promise it offered for the future of critical scholarship. But before we start, let us give some background.

The European Module in Spatial Development Planning has been an intensive, three-months-long, Erasmus accredited, residential course. Students were usually drawn from the approximately ten consortiated countries in Europe and moved to the university that hosted the programme for a period of at least three months. This framework earned it the title of *European Module* or EMSDP in short. The module started in Lille in 1997 as a joint PhD training initiative for spatial planning and development programmes, later extended to geography and urban and regional sociology. After having been based in Lille for several years (1997-2002), the module was run in Newcastle upon Tyne, hosted at the Global Urban Research Unit, School of Architecture, Planning and Landscape, Newcastle University (see Cameron's chapter in this book). In 2011, EMSDP moved to Leuven, managed by the Planning & Development Research Unit at the Department of Architecture. The objective of the module has always been to join up forces with academics and practitioners across disciplines and countries, to train postgraduate and PhD candidates in a collective way, using interactive methods of learning and tutoring, and rooting the training into an ethical stance, built on principles of equity, social innovation and sustainability.

Over a 20-year period (1997-2017), the module was delivered by a network of scholars (the ESDP network) which has been collaborating on various parallel research projects. In addition to joint research projects, the module has been a crucial emulsifier catalyst in the life of the network. It not only allowed regular contacts and cooperation between its members, but also became a way to assure continuous regeneration of the network. More than 200 students participated in its activities since its inception. In 2011, the EMSDP received the AESOP Association of European Schools of Planning Prize for teaching excellence.

As in our case, the class of 2006 was bubbling in multiculturalism with a cohort of students coming from nine countries, four continents, and a whole range of disciplines spanning from architecture, planning, landscape and design, to sociology, economics, business studies, psychology and forest engineering. Our diversity was striking, yet mutual affinities and deep curiosity, as well as a great deal of humour, ignited a catalytic loop that soon became self-generative. The methodology of the module facilitated the rise – during its running and for the following three-and-a-half years – of a deep commitment to action-learning.

Completely hooked by the intense emotional and intellectual support that the exchange was giving us – insights into different theories and approaches, training in critical thinking and questioning, overcoming isolation, and prioritizing *process* over *outcomes* – soon the already intensive module (with six class hours a day) wasn't enough anymore! We started to ask for ad-hoc lectures, we knocked at lecturers' doors, we kept discussing animatedly at every lunch time. Our diversity meant sometimes getting lost in translation, but shared values for inclusion and tolerance meant extra efforts in making sure that everyone took part, that we could communicate across cultures, and develop a deep self-awareness.

An important aspect of what happened that year was being exposed to the long-term academic collaboration of the teaching team, whose professional and personal ethics, as well as life-long intellectual project became a role model in many respects. We started to appreciate the powerful mixture of interdisciplinary and holistic approaches, blended with relations of trust and mutuality. The three months of full-day discussions, and particularly the task of presenting/reflecting on our own projects considering the module inputs, did effectively reshape, redirect, steer and/or variously sharpen our minds. For some of us, that experience signed a point of no return: it inspired a compelling desire for building collaborative journeys for social change. The ESDP network's leadership and personalized tutoring was crucial in shaping this experience. It provided us - at that time and up until now - ethical, intellectual and practical beacons to navigate the complexity of engaged scholarship.

Once the module was over, we explored ways to build our own social and intellectual project: we founded the EquaTerritoria network. Given the geographical, financial and career-related constraints that we experienced at the time, it is astonishing to realize the number of formative steps we went through (see figure below).

Perhaps the three most interesting of them were the following:

1. The establishment – by trial and error – of an organizational structure that, by coupling two people for each task (on a rota basis), ensured mutual learning, complementarity and efficacy. It was a way to meet our own learning needs and individual speeds alongside the desire to get the task done and meet the goals of our collective agenda;

2. The creation of a manifesto for the group, reflecting values and principles that informed our short and long-term objectives (see box below);

3. The development of a shared interdisciplinary methodology (matrix) for the analysis of case studies, reflecting the diversity of theoretical and disciplinary approaches of the group.

The active engagement in these - and the many other initiatives that constelled the life of the group - was an important learning trajectory, which has not only been crucial for our current life of activist scholars and dedicated practitioners, but also an opportunity to question our own pre-existing role-models, challenge the system we were supposed to go back to, reflect on the type of activism we wanted to engage with, and steer our academic trajectories in new directions. Twelve years on and that energy is still high: a number of us are still collaborating around teaching, research and activism.

A tipping point in the history of the EMSDP took place with the handing over of the baton from Frank Moulaert (the ESDP network coordinator) to a fresh new team headed by Pieter Van den Broeck (EquaTerritoria cofounder!) and Costanza Parra. As a result, in 2018 the EMSDP was diversified and reinvigorated under generous support from ESDP network, Planning & Development Research Unit, KU Leuven and the Flemish VLIR-UOS development cooperation agency. The newly titled Postgraduate Programme *International Module in Spatial Development Planning (IMSDP)* is wide-ranging, opening its doors to students from all over the world, with a particular facilitation for Global South colleagues. IMSDP is a blending of new and old minds, with the lifelong project shared by the people that created the EMSDP in the first place still very much alive, and shared as an intellectual, academic, social and political commitment to nurturing new generations of critical thinkers and doers. ◀

ACTIVITIES OF THE EquaTerritoria NETWORK

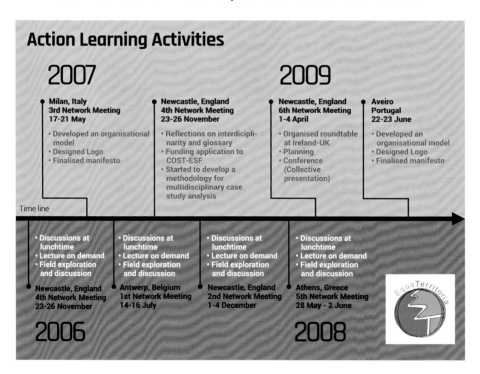

Action Learning Activities

2007

Milan, Italy
3rd Network Meeting
17-21 May

• Developed an organisational model
• Designed Logo
• Finalised manifesto

Newcastle, England
4th Network Meeting
23-26 November

• Reflections on interdiciplinarity and glossary
• Funding application to COST-ESF
• Started to develop a methodology for multidisciplinary case study analysis

2009

Newcastle, England
6th Network Meeting
1-4 April

• Organised roundtable at Ireland-UK
• Planning
• Conference (Collective presentation)

Aveiro
Portugal
22-23 June

• Developed an organisational model
• Designed Logo
• Finalised manifesto

Time line

• Discussions at lunchtime
• Lecture on demand
• Field exploration and discussion

• Discussions at lunchtime
• Lecture on demand
• Field exploration and discussion

• Discussions at lunchtime
• Lecture on demand
• Field exploration and discussion

• Discussions at lunchtime
• Lecture on demand
• Field exploration and discussion

Newcastle, England
4th Network Meeting
23-26 November

Antwerp, Belgium
1st Network Meeting
14-16 July

Newcastle, England
2nd Network Meeting
1-4 December

Athens, Greece
5th Network Meeting
28 May - 2 June

2006

2008

THE EQUATERRITORIA MANIFESTO

Manifesto – Final Draft

1. We, the signatories, agree herewith to establish a *multi-disciplinary network* dedicated to discussing and promulgating practices in all processes related to spaces for human living. We are thirteen researchers with different professional and ethnic backgrounds. We are trained in architecture, urban and landscape planning, engineering, psychology, economics, business, political science and sociology. We come from towns and cities across the globe, in Africa, Asia, Latin America and Europe.

2. We want to disseminate a *critical approach* towards spatial and developmental issues. We want to enrich and share our knowledge and practices from a multicultural perspective, stimulating our minds in questioning the world. The mix and richness of cultural and educational backgrounds, coming from and working in different inter-continental contexts are to be the strength of this network, enabling us to explore how different societies approach spatial and developmental issues, and how to deal with different strategies. This includes looking beyond an exclusively *Western* or *European* centred perspective in discussing issues, approaches and solutions.

3. We think it is important to *reconcile theory and practice* and contribute to the connection of the academic world with the domain of practitioners. We intend to influence prevailing discourses and link up with policy-makers and those who implement policies as we aspire to influence common practices. For this purpose, we are going to explore different analytical and practical approaches and tools, finding alternative ways to work within and outside our network.

4. We are a network with strong beliefs and values. *Sustainability and equity* are the main principles which inspire us and our work, because we believe human beings in all parts of the world deserve to lead a life in dignity. Therefore, the focus of planning has to be shaped to a more human scale. Spaces should reflect human needs rather than impose ideologies and theories.

5. We believe that space and places present *contextual specificities*. We want to raise awareness of place-sensitive approaches that include continuous dialogue with local communities, their knowledge and values. We share the belief that we must defend space/territories against commodification and indiscriminate standardization, by recognizing and valorizing their diversity. Therefore, we are not guided by values such as growth and profit. We question the capitalist model and recogniz the need to search and develop alternative development models.

6. We are critical about the use of *language*. We maintain the need to constantly analyse and unfold language and concepts to avoid an uncritical reproduction of existing models and discourses. We reject current neoliberal rhetoric, which appropriates the discourse of globalization as a means to impose a dominant ideology and practices. We question this rhetorical hegemony - replacing neoliberalism with globalization - and want to explore how the evident positive aspects of globalization can support our work and ambitions.

7. We believe in *diversity and pluralism* and want to initiate a *broad and open dialogue*, but this does not imply a dilution of our stated values. It does, however, have consequences for the way we work together. First of all, we understand that terms, concepts and issues can be discussed from multiple perspectives and that in different parts of the world this can lead to different interpretations. While it is not our task to reconcile these contradictions, we will aim to recognize and acknowledge them. While sharing ideas and solutions, we adopt a tone that recognizes that values may evolve and change through discussion or changing contexts. We maintain the need to start from local potential and knowledge, as we perceive planning to be a participatory process that recognizes bottom-up initiatives without rejecting top-down approaches.

We invite everybody who shares these values and vision to join us and to contribute to the activities of the network.

Newcastle upon Tyne, 1 December 2006, revised 22 June 2007
Antika Sawadsri, Barbara Van Dyck, Brenda Galvan-Lopez, Chiara Tornaghi, Daniela Coimbra de Souza, Elena Bertè, Elisabeth Brooks, Giancarlo Cotella, Giota Karametou, Jérôme Hassler, Norn Rittirong, Pieter Van den Broeck, Rosanna Grasso, Tamer Ahmed (EMSDP 2006) ◄

33. SOCIAL INNOVATION AND STRATEGIC SPATIAL PLANNING PEDAGOGICALLY INTERACTING IN LEUVEN

Loris Servillo

Although the European Module in Spatial Development Planning (EMSDP, see chapter Tornaghi et al.) started being organized at the KU Leuven, it affected the urbanism and planning curriculum in the Department of Architecture, through inputs of the EMSDP programme and ESDP network teachers in the courses of *Strategic Spatial Planning* (SSP) *and Institutional Aspects of Spatial Planning* (IASP). The courses aimed to deconstruct mainstream readings of our society in space and of the role that spatial planning could play in it, revealing the complexity of socio-spatial dynamics, the contradictions of neoliberal regimes in urban and regional dynamics (Moulaert et al., 2010), and the articulated interconnections that affect socio-ecological systems. Activities of the teachers, including members of the ESDP network, Frank Moulaert and myself, were oriented to awakening the critical capacity of students in reading space as socially constructed (Massey, 2005; Khan et al., 2013), and to providing them with the cognitive tools to deconstruct existing practices in planning and create counter narratives.

In eight years, the SSP and IASP courses have evolved, mirroring a desire to reflect on contemporary challenges. In particular, it is within the SSP course that teachers mobilized their extensive corpus of knowledge and pedagogical beliefs (Moulaert et al., 2010; Martinelli et al., 2013) to extend exploration of the technical and procedural dimensions of strategic spatial planning - the ways in which strategic spatial planning is different from traditional statutory planning - to a larger and critical view on the role of planning in society and how a strategic approach can better meet the challenges that characterize our time (Moulaert et al., 2013b; Albrechts et al., 2017). This evolution represents the specific significance of the European Module for the KU Leuven's Master programmes in Human Settlements and Urbanism and Strategic Planning.

Along these years, the focus of the SSP course has become more linked to the goal of sustainable development (SD). This evolution has happened in full awareness that, as for many contemporary concepts, there is an abundance of meanings of *sustainable development*, floating around in ideological and policy debates, as much as in scientific analysis. At the same time, there is a certain disappointment in the way sustainable development is taken into account in strategic and spatial planning projects and concepts, especially a prevalence of "soft" approaches that are more in tune with the contemporary conjuncture of neoliberalism and real-estate hunger.

Therefore, the course dedicated progressively more time to providing a stronger elucidation of sustainable development and its ethical, normative and analytical dimensions. In particular, the course connected the challenges and struggles that planning is facing worldwide to the conceptualization of socio-ecological developments (Parra and Moulaert, 2016). In doing so, it located the planning practices within the larger frame of governing socio-ecologically connected elements, stretching the implications of pursuing sustainability and (ecological, social and spatial) justice through planning. On the one hand, this contributed to a crucial shift toward holism (Moulaert and Mehmood, 2014) as an interpretative approach that advocates the effort of understanding complexity and interlinked relationships between humans and the environment in which we always act to reproduce space. On the other hand, this search for deep understanding of sustainable development implications aimed at strengthening the social dimension of sustainability to foster social cohesion and democracy (Parra and Moulaert, 2010).

In line with the latter, the course extended the theory and the concepts of SSP as a technique in the planning domain with a stronger focus on socially innovative practices. Beside the approach to strategic planning as a practice to be identified in regional planning, urban plans and projects, the course tightened the relationship with social innovation, opening up opportunities for a wider reflection on insurgent planning actions and alternative forms of bottom-up strategic thinking that counteract hegemonic practices. The capacity of socio-spatial actions to enable the expression of needs from unheard groups of people and to establish new values and uses of space has been brought to the forefront.

Complementarily, the urge for deeper exploration of contemporary challenges had a deep impact on the IASP course as well. This course changed from its initial focus. It went from being mainly dedicated to formal institutional dimensions, strictly related to national and regional planning systems, to a wider view on the institutional dimensions of planning (Servillo and Van den Broeck, 2012), looking at composition of arenas, procedural dimensions, and power relations (Servillo and Lingua, 2014; Munteanu and Servillo, 2014). It reflected on how planning practices change when faced with specific socio-economic challenges, bringing in or avoiding discourses and themes in planning agendas. As such, it offered an extensive variety of theoretical and practical investigations, with a world-wide perspective.

The pedagogical arrangement of the two courses offered to students challenging – and sometimes disturbing – views on the planning dimensions. They complementarily deepened the focus on how different socio-institutional dimensions can have great influence in triggering deep strategic approaches to space and its transformation (Moulaert et al., 2016a; Servillo, 2017). At the same time, the students brought into the discussion international cases which were outstanding examples of either innovative planning experiences that brought emancipation and empowerment of marginalized groups, or power dominant exploitation, as a kaleidoscopic representation of socio-spatial dynamics and related power struggles in the shaping of cities and socio-spatial processes in the world.

This kaleidoscope, made of theories and practices investigated in the two courses, constituted a powerful laboratory which enabled students and lecturers to deconstruct and reread contemporary spatial dynamics. As such, SSP and IASP teachers have explored the rabbit hole with passion, in the hope that it made students wiser protagonists in the world of today and tomorrow. ◄

13

BRIDGING SOCIETY AND ECOLOGY

Bosque de Alerces, Chiloé Continental © Constanza Parra

34. BRINGING THE SOCIAL BACK IN SUSTAINABLE SOCIO-ECOLOGICAL DEVELOPMENT

◆

Constanza Parra, Angeliki Paidakaki,
Abid Mehmood and Pieter Van den Broeck

At the turn of the new millennium, Moulaert et al. published the book *Globalization and Integrated Area Development in European Cities* (Moulaert, 2000). One major conclusion discussed in this book was the significance of 'grassroots democracy and the empowerment of local communities in delivering a social, economic, and cultural renaissance which meets the needs of the local population more effectively than the market-forces creed' (Moulaert et al., 2010: back cover). Social innovation was in this book portrayed as a condition sine qua non of alternative local development (Moulaert, 2000: 66). This book and other contributions on social innovation and territorial development were generous in offering a variety of perspectives from which the decline and renaissance of neighbourhoods and cities could be examined (see other contributions to this book). We noted that, among all these perspectives, nature, environment, and sustainability concerns were being relegated to the backdrop.

We start this essay by referring to the book *Globalization and Integrated Area Development in European Cities* because it contains first hints connecting the problématique of social innovation and community development to socio-ecological concerns. The importance of a (sustainable) relationship between humans and nature is expressed in the following quote of the book: 'The integration of the dialectics between nature and society (ecological processes and planning) also plays a part. Nature is to be considered as the most primary resource of social progress and should be preserved and reproduced according to this status' (Moulaert, 2000: 43). This is worth noting as it connects nature to a definition of progress that is essentially social, and also highly respectful of nature and ecological processes. From this point of view, social innovation in territorial development is directly linked with an ethic of collective action that is socially sustainable and respectful of nature-culture entanglements (Parra and Moulaert, 2011).

The above quote also anticipates a socio-ecological shift in social innovation literature which led to the construction of a bridge from social innovation to sustainable socio-ecological development. As such, Moulaert's work on social innovation in territorial development has become foundational to new social science research touching on a variety of environmental and sustainability themes that have gained prominence in the last few years. These works include doctoral theses on subjects such as water conflicts and decision-making (Calvo Mendieta, 2005), socially innovative sustainability in protected areas (Parra, 2010), social innovation and urban sanitation (Putri, 2014), social resilience cells in post-disaster places (Paidakaki, 2017) and hybrid governance in urban gardening (Manganelli, 2019). Several publications on social sustainability, social innovation in and for sustainable development, governance of socio-ecological systems, resilience and housing, and governance of the landed commons are examples of this socio-ecological turn.

In this essay, we cover the question about how knowledge on governance and social innovation have become part and parcel of the study of socio-ecological development. We believe that this is a relevant question as it represents the connection between the work carried out by two, or perhaps three, different generations of researchers. In the remainder of this essay we proceed in three steps. We start by recalling the contribution of Old Institutionalism and how this inspired work on territorial development and social sustainability. Second, we look at current work aiming at rescuing the social, political and cultural components of sustainability, resilience and socio-ecological development. Third, we look at the future, with optimism, and figure out socially innovative research agendas and socio-political concerns that we hope we will continue researching.

AN OLD INSTITUTIONALISM INSPIRED SOCIAL SUSTAINABILITY

In his time as Professor of Economics at the Université de Sciences et Technologies de Lille - Lille 1, Moulaert used to teach the courses *Institutional Economics* and *Firms and Institutions*. In these courses students would engage with the works of Commons, Polanyi, Weber, Boyer, Hodgson... about the existence of something called heterodox economics, evolutionary economics, holism, pragmatic American philosophy and the fallacy of the centrality of markets in regulating our *vivre ensemble*. Institutionalism and regional development theories, inspired in Old Institutionalism scholarship, were an important first step in the journey to approach social sciences to sustainable development and the natural environment (Parra, 2010; see Calvo Mendieta in this volume).

Several normative principles concerning socio-ecological systems and their sustainability were brought together at the service of a socio-institutionally inspired analysis of the inherent *social* nature of the economy and development (Söderbaum, 2000). These principles highlighted, for example, the plurality and intrinsic values of nature, and called for an approach to territorial development connected to the sustainability of ecosystems and their precious status as primary resource for societal development. This approach contains a criticism of the instrumentalization and monetarization of the environment which narrows the questions on values to mere economic standards. Instead, values are seen in a comprehensive way, by combining aesthetic, cultural, social, political and historical significances and their interactions. Close to this perspective comes the work developed by Polanyi on the artificial subordination of society, social relations, labour, land/nature and money to market logics. According to Polanyi (1944/2001), before the nineteenth century the human economy was embedded in society and therefore subordinated to social relations, politics and religion. After that, classical economists tried to create an economy detached from society yet despite their efforts they could fortunately not achieve this goal (Block, 2001: xxiii). For Polanyi, it is a big mistake to treat nature and human beings as objects that can be traded in a market. Nature has a sacred intrinsic value which cannot be denied (Block, 2001: xxv).

BRINGING THE SOCIAL BACK

Bringing the social back to the fore has been the primary mission in our collective work on sustainable development, resilience and socio-ecological systems. By social we mean societal concerns, social relations within development, and political, cultural, governance and ethical ingredients contained in the different types of relationships between human beings and nature (Parra, 2013). This mission, aiming at highlighting and reinforcing the societal character of the environmental problematic, has taken different forms. One of the first attempts was in the concepts of social sustainability and socially innovative sustainability. This was done conceptually, starting from old institutionalism and economic sociology, as well as through the empirical analysis of protected areas for nature conservation. Parra and Moulaert (2010, 2011) criticized the predominance of the economic and ecological sustainability dimensions which deny sustainable development of its social and human distinctiveness. The aim was to give a robust status to the social pillar, enhancing its classic interpretation, revolving around equity and fairness, to the issues of democracy, governance and social relations. This work introduced the *process* dimension of social sustainability: the *social* in sustainable development is materialized in terms of governance as a socio-political, multi-partner and multi-scalar process, as is also the case in social innovation. This means that social relations and collective action becomes the core driver of, and a condition for, social sustainability. Sustainability is therefore the dynamic outcome of social processes, in which factors such as power relations, conflict, socio-institutional arrangements, territorial specificities and local culture play key roles.

The study of protected areas in the French context was enlightening in this conceptualization of social sustainability and its connection with social innovation, as shown in the following quote referring to the Parc Naturel Régional du Morvan case:

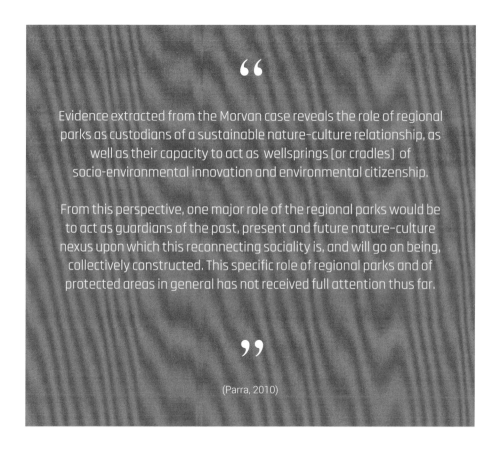

"

Evidence extracted from the Morvan case reveals the role of regional parks as custodians of a sustainable nature-culture relationship, as well as their capacity to act as wellsprings [or cradles] of socio-environmental innovation and environmental citizenship.

From this perspective, one major role of the regional parks would be to act as guardians of the past, present and future nature-culture nexus upon which this reconnecting sociality is, and will go on being, collectively constructed. This specific role of regional parks and of protected areas in general has not received full attention thus far.

"

(Parra, 2010)

From these analyses, the concept of social innovation was reinterpreted in the light of the sustainable development problematic, leading to the concept of socially innovative sustainability that refers to *a collective definition of sustainable paths of development, innovation in the governance for sustainable development* and *enhancement of environmental rights – as a basis for new environmental citizenship rights* (Parra, 2010).

Bringing the social back has also been done through the work on *social resilience* and *social resilience cells*. Using the example of housing systems in New Orleans and examining the interactions between discursive hegemonies, material practices and institutional arrangements overplayed in post-disaster contexts, Paidakaki and Moulaert (2017, 2018) have revisited the notion of resilience, attaching to it stronger socio-political features. The governance concept of Social Resilience Cells (SRCs) was developed with the aim of highlighting the heterogeneity of resilience qualities and stressing the role of power relations in guiding the implementation of a city's "resilience". SRCs are explained as housing policy implementers, both for-profit and non-profit, who have their own language and defend their discursive and material practices based on their own common values, needs, and aspirations. Some SRCs are more powerful compared to others in terms of recognition, access to resources, and facilitation of their needs. This asymmetry to power relations ultimately results in some resilience actions being promoted and others being wasted away. When an urban system is produced, co-produced and hetero-produced according to the divergent demands and qualities of various SRCs, then a more socially-just post-disaster governance modality should be pursued aiming for the redistribution of resources and cultivation of empowerment of all SRCs struggling for their right to the reconstruction experiment.

The governance of socio-ecological systems is another research theme that caught our attention. In a workshop organized by Leuven Space and Society in 2016, Moulaert, Van Dyck and Parra (2016) questioned the relevance of the socio-ecological systems approach propagated by the Planetary Boundaries framework. The framework is deemed too rationalistic and socially disembedded, over-relying on quantitative-techno managerial approaches and modelling, and keeping distance from social justice aspects in a society. Governance of socio-ecological systems is not simply about designing new modes of governing and setting the rules "right", Moulaert, Van Dyck and Parra (ibid) argue. On the opposite, governance systems have a history and geography; they are spatially situated, culturally moulded, and embedded in social structures fragmented along various conflict lines. To that end, the authors develop a compelling argument that governance should be analyzed within its socio-ecological dynamics (Mehmood and Parra, 2013). This translates into a socio-spatially situated and dynamic understanding of governance wherein power, path-dependencies, social diversity, culture, reproduction of collective frame-of-minds, symbols and institutional practices, play a key role, and become part of the overall socio-ecological reproduction dynamics. The analysis of the San Pedro de Atacama case-study in Chile offers interesting insights on the governance of the nature-culture nexus by highlighting the entangled social, political, cultural and ecological dimensions of socio-ecological development. Parra and Moulaert (2016)

Morvan Regional Park, Bourgogne © Constanza Parra

Pumalin Park, Chiloé Continental © Constanza Parra

subsequently offer three powerful messages derived from the study of this Chilean desert locality. First, culture is not there to be governed but is also a constituent of the governance systems. It is people that govern with the means reproduced in and drawn from socio-ecological systems. Second, governance is an integral part of socio-ecological metabolisms in which contradictory processes and conflictive agencies take place (Swyngedouw, 2005). Third, the governance of the commons is consequently a political process rather than a rational process of (collective) management procedures (Parra and Moulaert, 2016).

SOCIAL INNOVATION FOR A MORE SUSTAINABLE AND JUST SOCIO-ECOLOGICAL DEVELOPMENT

It is not a coincidence that this essay on the social aspects of the environmental and sustainability problematic has been written by four authors working in close collaboration with Moulaert. Social innovation in and for a more sustainable, democratic and just socio-ecological development is at the core of the research and action agenda that we – as believers – would like to follow in the years to come, to seek more harmonious human-nature relationships and pursue for a socially just and sustainable world (Mehmood and Parra, 2013). Our attempts to further consolidate this socio-ecological turn in the analysis of territorial development will continue at least in the following directions:

• The role of social innovation as action research agenda in response to the problems of unsustainable practices and unsatisfied social needs while pursuing the challenges of environmental degradation, emphasizing the socially transformative capacity for enhanced and creative sustainability.

• The socio-political nature of the governance of socio-ecological development.

• (Old) institutionalist approaches to sustainable territorial development, including among others Polanyi's work on fictitious commodities such as nature and land.

• Gender, ethics and diversity as core themes for a progressive research agenda on sustainable development, resilience, political ecology and social innovation.

• Alternative forms of our *vivre ensemble* comprising, for example, the resurgence of the commons such as those traditional territorialized commons assembling forests, agrosystems, water and the entire diversity of ecosystems and cultural systems.

So, as it appears, bringing back the social in sustainable socio-ecological development is a complex task that requires critical reflection, collective effort, and a whole lot of belief, to overcome the inertia of unsustainable practices and market-driven ideologies, and to promote measures that combine ecological, social, institutional and political action. ◄

35. 'THEORIE DES CITÉS' AND WATER CONFLICTS

◆

Iratxe Calvo Mendieta

Water use conflicts, in neoclassical economics, are seen as a situation where resources are not optimally allocated due to a *market failure*. Following this conceptual standpoint, conflicts can be resolved by simply repairing the information gap between users, most of the time by adapting prices. And that is what most students of economics are taught. To understand water conflicts, the *Théorie des Cités*, proposed by Boltanski & Thévenot (1991), offers an interesting perspective on the origin of conflicts and essential components of conflict dynamics. According to this framework, individuals in a society refer to six basic models when justifying and legitimizing their behaviour (the *cités*): inspired, domestic, opinion, civic, industrial and commercial. Each *cité* enables the interpretation and the coordination of actions but they often contradict each other. Indeed, the behaviour of individuals engaged in a conflict depends largely on the representational systems they refer to. When different stakeholders are in conflict and arguments are defended, individuals detach themselves from their situations to achieve more legitimacy. This architecture of the *cités* and their dynamics enables better reading of conflict situations: individuals are not only influenced by prices or standards, but also by representational systems. The *cités* help us to better understand how, in a particular situation of collective action, some individuals get into conflict and others cooperate, and to avoid the potential crisis created by the confrontation of different representational systems, or, turn them into innovative dynamics and socially innovative initiatives. Besides violence and power struggles, individuals can choose to engage themselves in three aspects: local arrangements, relativism and compromise. Among them, compromise is the option that allows conflict resolution.◄

36. UNDERSTANDING LAND USE RIGHTS, BUILDING LANDED COMMONS

---◆---

Pieter Van den Broeck

Responding to a new wave of privatization, market fundamentalism and land grabbing (*enclosure*) since the 1980s, academics, civil society activists, social movements and public intellectuals have in the last two decades put the concept of *commons* back on the agenda. The work on how socially innovative initiatives succeed in transforming social relations and empowering underprivileged actors is also about the transformation of predominant market-based mechanisms commodifying land and labour, with the aim of contributing to a solidarity-based society. The different strands in the work on social innovation, territorial development, regulation theory, Keynesian economics and political economy in general thus contribute to a better understanding of the socio-economic mechanisms of enclosure as well as the ways in which they can be reversed into mechanisms of *de-closure* or *commoning* (Van den Broeck et al., forthcoming). More recent focus considers the nature-culture nexus, the governance of socio-ecological systems and especially on the understanding of biophysical and social processes as mutually constitutive. This work not only explores political ecology, post-structural approaches, rural development and sustainability literature but also the literature on commons. Wider views look on land tenure through the real estate mechanisms, globalization and urban development in general, rent extraction, the role of informality, community land trusts and housing for the urban poor. These tracks culminated in the research project INDIGO on land tenure, land policies and commons, focussing on innovative commoning practices and the ways they oppose predominant mechanisms of enclosure (www.theindigoproject.be; www.buildinglandedcommons.be). The project stresses the need to move from land ownership and property to a governance perspective in which land tenure is positioned in its institutional dynamics, a broad understanding of commons - epitomised in ten principles of the landed commons - and a focus on how commons are actually built, the latter being enabled by mobilizing and adapting the *Théorie des cités* on how compromises are negotiated. The project leads to a methodological guide for the analysis and planning of landed commons. ◄

© Alessandra Manganelli

© Alessandra Manganelli

37. LAND USE BETWEEN SPACES AND TIMES

Hubert Gulinck

Given the rigidities of land use nomenclatures, a socio-ecological approach helps in seeking a new epistemology of land use. There are old and new realities, and environmental and societal arguments to increase the attention to more fluid land use approaches, including transitional categories in space and in time. Examples include urban agriculture, carbon sequestration, greenspaces, brownfields, and many others. Innovative generic interface types can be proposed as a framework to accommodate this heterogeneous set of conditions at the spatial, temporal and cultural margins of customary land use categories: *hybrids, guests, gardens, buffers, overuses, fallows, commons, connectors, edges, reaches, and residuals* (Dewaelheyns et al., 2011).

These are not exclusive types, but rather semantic overlays of transitional conditions. An example is the introduction of beekeeping in much neglected river buffers in Tanzania that rapidly induced spontaneous vegetation restoration, increased water quality and quantity, improved pollination, and provided a boost to the local economy, community self-confidence, rural development and human health (Kimaro et al., 2014). The deficient river shores turned into effective *buffers*, and into multifunctional *hybrids* with a wide *reach* of positive impacts, so ultimately margins became focus. Such cases support the need to link socio-economic, socio-political and socio-cultural perspectives with environmental planning. This includes looking at the interface categories inspiring land use planning and governance in a wide range of contexts such as peri-urban areas, the agriculture-conservation nexus, and the incorporation of ecosystem and socio-cultural services in planning and development models. ◄

14

OYONG

MANG

INTERNATIONAL
DEVELOPMENT

38. LEARNING FROM COMMUNITY DYNAMICS IN KENYA'S 'INFORMAL' LIVELIHOOD AND HOUSING STRATEGIES

———————◆———————

Emmanuel Midheme and Margaret Macharia

The following cases, from Eastleigh and Voi, highlight how the outcomes of community initiatives and negotiated access to livelihoods, as well as shelter, reflect the struggles that groups, with no access to capital-endowed networks, experience in their endeavours to live decent lives within Kenya's rapidly transforming cities.

Empowerment and exclusion mechanisms often operate alongside each other in many cities of the Global South. Within Kenya's rapidly transforming cities, community initiatives - in the form of *informal* livelihood and housing strategies - have been potent tools employed by marginalized groups to challenge enduring forces of socio-spatial exclusion. This shows how community mobilization and initiatives work best if they are supported by larger networks (Moulaert et al., 2014).

In Nairobi's Eastleigh neighbourhood, socio-economic empowerment processes amongst the predominantly Somali trading community have been made possible largely through the mobilization of both local and transnational socio-cultural clan networks. In addition, other instruments such as the *hawala* money transfer system, which facilitates long-distance trade amongst migrant Somalis in the diaspora, has greatly supported the traders' economic empowerment. Nonetheless, both Somali and non-Somali street vendors in Eastleigh maintain their struggle for urban inclusion through appropriating, for their trading activities, the residual road-side spaces along Eastleigh's First Avenue and the front pavements of the ubiquitous *shopping malls* in the area. The vendors' ability to maintain their trading slots on the street depends on constant negotiations between the small-scale traders, on the one hand, and the shopping mall owners and the local authorities on the other (Macharia and Van den Broeck, 2016).

A similar struggle - this time for housing - by a marginalized squatter community in Voi has given rise to a novel model of landholding and housing ownership through a Community Land Trust (CLT). The Tanzania-Bondeni Community Land Trust - a culmination of lengthy negotiations amongst the local community, state agencies and private landowners - is a radical departure from the hitherto prevailing land management models in Kenya that have, since colonialism, emphasized individualized land ownership at the expense of communal landholding. Despite the struggles and challenges of implementation, the Voi CLT has withstood the forces of land speculation, enabling residents to own land and housing under terms that would otherwise be impossible in the open market (Midheme and Moulaert, 2013). ◄

Eastleigh's First Avenue: Shopping Malls vis-à-vis Street Appropriations by Vendors 2011 © M. Macharia

39. PUSH IN THE KAMPUNG: SOCIAL INNOVATION IN THE GLOBAL SOUTH

◆

Prathiwi Widyatmi Putri

More than two decades of research have significantly contributed to the debates on neighbourhood development and community empowerment, providing insights beyond the economic and technological readings on social innovation (Moulaert, 2009; Moulaert and Jessop, 2013; Moulaert and Leontidou, 1994; Moulaert et al., 2001b). Several of the works stress the roles of place-based and multiscale-connected governance structures involved in different forms of collective actions within or outside the spheres of state and market (among others, Moulaert, 2009; Moulaert et al., 2010). This offers analytical promise for understanding diverse urban contexts in the Global South to address the reality of extreme poverty, political marginalization in spatial development and asymmetric policy making processes, and to promote community aspirations and collective practices.

My own interest in a socio-ecological approach to social innovation in the Global South was inspired from such works, to empirically demonstrate the theoretical and methodological significance for Jakarta's kampungs. The case of kampungs exemplify local spatial entities of cities in the Global South, characterized by different facets of informality that are operating as collective networks fulfilling basic needs of the people. Despite their often-deprived sanitary conditions, which is the common rationalization used to evict the communities, different ingredients of social innovation could be found in Jakarta's kampungs. These include diverse cultural identities and socio-ecological practices that provide alternatives to the homogenizing consumption patterns of urban economic systems, bottom-up tailored initiatives from broader civil society to overcome social-spatial exclusions, and alternative models to the large scale urban development projects. While most kampungs have been excluded, particularly from the state water and sewerage networks, different actions have been initiated for water and sanitation service provisions by means of regularizing networks of informal relations and/or introducing new infrastructure systems. A joint NGO-community initiative in Kampung Kojan is one such example, explained below.

Mercy Corps worked in Kampung Kojan, Kalideres sub-district, from May 2009 to December 2010. Through its Program of Urban Sanitation and Hygiene Promotion (PUSH), it delivered 219 technologically innovative septic tanks for treating black water. The aim was to test and promote an alternative technology of low-cost modular watertight septic tanks that meet the quality standards set by the capital territory (DKI Jakarta). In all, thirty-five shared latrine units were built for tenants of rental rooms, in addition to a unit of public toilets and washing place. This was a response to the fact that many Betawi people in the kampung have built rental rooms on land that was once used for subsistence agriculture. This small scale housing provision meets the high demand for cheap housing for the workers and helps sustain the household income structures of the community. But providing toilets and facilities for bathing or washing was often neither their interest nor a necessary element within their business cycles. The strategy of providing communal and public facilities has addressed the needs of temporary workers and transient migrants, who can be considered among the poorest groups in the Kampung community. The tenants enjoyed immediate benefits of having latrines near their rental rooms, saving the time for domestic work and hence increasing the value of their income. Shared septic tanks proved more efficient, and increased the quality of shared open space between rooms and houses in the crowded settlement. As a future strategy to anticipate increased/increasing affordability and growing preferences for privacy, communal septic tanks would be available in the Kampung to collect the faecal waste from newly built individual latrines.

The PUSH project approach was socially innovative because it applied two pillars of sustained and meaningful local development: institutional innovation and socio-economic innovation, that is the satisfaction of various basic needs of local communities (Moulaert et al., 2010; Moulaert and Nussbaumer, 2005). Institutional innovation helps channel cultural emancipation, interpersonal and intergroup communication, and preference revealing within decision-making mechanisms (Moulaert and Nussbaumer, 2005). PUSH allowed an accumulation of local knowledge about public health and environmental sanitation as well as a better understanding of existing collective practices in meeting water and sanitation needs. The project stimulated innovation in social-spatial relations, both within and beyond the neighbourhood level. In some kampungs, Mercy Corps also initiated community enterprises to collect the sludge from septic tanks using small vehicles equipped with a low capacity pumping system. This follow-up project added an important element to the social infrastructure, one that could become the seedbed for collective wastewater management across other kampungs.

Despite its potential, the PUSH project has not been able to create and institutionalize spaces for long-term interactions and commitments between communities and urban professionals in several fields, including government officials, activists and development aid managers. There are two key reasons for this failure. First, not enough time has been allowed for the project outcomes to develop. Second, continuing kampung evictions in Jakarta tend to prevent the establishment of new innovative agencies in response to ill performing state water and sanitation service provisions. ◄

© Prathiwi Wydiatmi Putri

Allah Wasai CRP — Shop at Home © Tahira Tarique

40. 'BUSINESS-IN-A-BOX' AS TRANSFORMATIVE SOCIAL INNOVATION

◆

Abdur Rehman Cheema
and Abid Mehmood

B usiness-in-a-Box (BiB) is part of a community-based development programme in rural Pakistan which has led to some transformative and capacity building actions. Initiated by the non-governmental Rural Support Programme Network, the model closely aligns with the approach relating social innovation for human and community development to three key aspects: satisfaction of basic human needs; improving social relations; and, socio-political empowerment (Moulaert et al., 2005, 2013b).

Implemented in two pilot phases between 2013 and 2016 in three rural districts of

Pakistan, the programme comprised of two parallel actions: a micro-franchising setup through training of local women as Community Resource Persons (CRPs); and, a local capacity building project through Village Health Committees (VHC).

• The micro-franchise initiative trained 450 women CRPs to assess community needs and potential for health-related products and sell a BiB kit containing daily household usage items, health commodities and short-term contraceptive methods through household visits. Initial results have shown success with 84% of CRPs who were willing to continue working with BiB after the project since it provided them with a reliable source of income and reduced their dependence on the male family members.

• Village Health Committees (VHC), as institutionalized vehicles comprising of a mixed-gender group of key actors and stakeholders was a useful medium for recruiting the CRPs, negotiating the provision of preventive and reproductive health related products with households addressing the social and cultural taboos in the use of such methods at the community level. Subsequently, the VHC played a key role in managing supply chains and broadening their interactions with other tiers of local, district and provincial government to mediate further support for other health related matters in the remote villages.

Still in its infancy, the model has demonstrated the potential to scale up, while more refinement is required to address the challenges of horizontal and vertical collaborations. As a model of place-based social innovation it can be implemented in other sectors and geographical settings to avail positive behavioural changes along with financial sustainability and assistance in refining the approaches to rural health provision, accessibility and policy besides reducing the burden on public health services. ◄

© Maarten Loopmans

15

WAYS FORWARD

167

41. WHEN INNOVATION LOST ITS SOCIAL CHARACTER, OR NOT?

Frank Moulaert

S tarting from the history of thought and practice of innovation, I will muse on the trajectories of *meanings* and *theorization* of this concept. I will use this history as a mirror to reflect on the relationship between innovation and development. I will explain how originally, say starting in the eighteenth century, the term innovation basically referred to *social* innovation and referred to debates and struggles to change the world (Godin, 2015). But with the rise of modernity and the strong appeal of science and technology to policymakers and other societal and economic leaders, the concept of innovation lost its essentially social meaning. As of the 1930s, innovation has been predominantly referred to as technological innovation, connecting it preferably to economic innovation. This reduction of the meaning of innovation has had an impact on the way development was conceived and materialized, especially at the expense of visions and strategies of development from below.

In this short reflection on the relationship between innovation and development, I will mainly dwell on three questions:

- What has been the social character of innovation and how did it get lost?

- How to reconnect innovation with its social roots?

- And how this reconnection can lead to a more integrated approach to the relationship between innovation and development in its various human dimensions?

INTRODUCTION

This book is a deep collective reflection on the relationship between different types of socially innovative strategies and processes of human development through territorial dynamics embedded in an interscalar world which is still dominated by capital accumulation, political power for the few and socio-cultural exclusion; yet a world full of opportunities to do things differently through social innovation in development strategies and governance. The contributors have brought back to our collective minds the shared trajectories along which, through a diversity of co-produced methodologies, brave attempts have been made to relate social science analysis of agencies of development to actual development processes, especially in urban and regional contexts. Core concepts in the study of territorially embedded development are *innovation* and *(spatial) systems of innovation*. Unfortunately, over the last century, the relationship between innovation and development has digressed from its original natural mutuality. This short text intrudes into this digression and examines how *innovation can retrieve its social character*. The following section deals with some critical issues in the theory of innovation and development, and it raises some critical issues in the interpretation of innovation within the context of socio-technical systems (Jessop et al., 2013). *How humane is the conception of innovation in such analysis?* is the main question that is addressed there. The historical "brush up" in the subsequent sections reminds us of the social roots of innovation to help reconsider the relationship between innovation and development, by re-socializing innovation as one among diverse human strategic agencies. The final section draws some lessons for social science emerging from a revisited relationship between innovation and development.

CRITICAL ISSUES IN THE THEORY OF INNOVATION AND DEVELOPMENT

The ambition of the paper is not to retrace the history of innovation theory and innovation systems, but solely to identify some milestones in the understanding and analysis of innovation in connection to development and the role of other human agencies in development dynamics. Important milestones are the de-socialization of the concept of innovation, the rise and predominance of technological innovation and its embedment in market-oriented innovation systems, as well as the return of social innovation social science analysis and a gradual rise of alternative interpretations over the role of innovation in development. Important in this respect is the concept of socio-technical system and how it has been criticized and reconstructed over the years (Moulaert and Swyngedouw, 1992; Smith and Stirling, 2010). Socio-technical systems are a core concept in science and technology policy and analysis, but their definition is not unambiguous. Till today the predominant definition keeps the institutional dynamics subordinate to market and technological logic. National, Regional or Local Innovation Systems are the more concrete constituents of a socio-technical system. Fortunately, recent readings of Social Technical Systems have made technology more interdependent with institutional dynamics and socio-technical regime change (see for example Smith and Stirling, 2010). At the 2017 Globelics Conference in Athens (11-13 October), Lundvall stated in his keynote address that

'*Science and technology* have much to offer to solve the problems of mankind, if the *institutional settings* allow people to have the *capacity to learn* not only skills but also new norms and values.' (freely paraphrased).

Next, he wondered

'What kind of institutional change is necessary to foster *new technological trajectories* that give priority to *ecological sustainability* and the creation of better *living conditions* in the poor countries.'

This expresses a real concern about the direction technological trajectories take and the conditions of development they can generate. It also expresses a concern about the nature of institutions, how they are built and how they affect development. In most studies of innovation systems and socio-technical systems, the nature of innovation is not questioned and the relation to other types of human agency not considered. These relations imply a diversity of institutions and institutionalization processes. Evolutionary and institutional economics, as well as science and technology analysis did a good bit in unravelling the institutional complexity facilitating or hampering technological innovation, stressing that it cannot be studied unless it is considered embedded in a *socio-technical system*. Still this embeddedness approach misses a large part of the social complexity of real life innovation and development dynamics.

I identify five shortcomings in mainstream socio-technical systems approaches and the way "their" innovation systems are conceptualized:

- The relationship between institutions and technological change is often conceived in an instrumentalist way, with a causal relationship going from institutions to technological change. Technological change is enabled by existing, transforming or newly created institutions. Institutions are in most socio-technical systems considered as institutionalized agencies making innovation happen (Moulaert and Jessop, 2013; Moulaert et al., 2016a). These socio-technical systems exist within a variegated macro-institutional system that as a whole should facilitate technological innovation.

- In most socio-technical systems and their innovation systems social dynamics are reduced to institutional dynamics. Yet in real life a socio-technical system is built on a diversity of social relations, referring to the different spheres of social life, from the domestic, through educational and cultural, to the socio-political sphere. These social relations may also attain some form of institutionalization; but their causal orientation, if any, will only coincidentally be related to technological progress. These relationships reflect yet other parts of human life than those at play in technological progress (Moulaert, 1996).

- Socio-technical and innovation systems analysis tends to underplay the role of power relations in institutional dynamics and the actual directions which innovation trajectories often take (Arocena and Sutz, 2003).

- Focusing in a very rational way on the materialization of new technologies and an institutional system that will support them, socio-technical systems hardly account for diversity and outlier behaviour that do not fit the classical R&D picture. We know from reality that innovation goes over stumbling blocks and often through phases of informal socialization before it becomes established or … disappears.

- Although there has been an evolution in the way innovation as an agency and a process has been defined, in the innovation systems literature the connections between technological, economic, institutional, organizational and social innovation have not been studied in depth (Moulaert and Hamdouch, 2006; Hamdouch and Moulaert, 2006). Institutional innovation is understood as innovations in institutions that are beneficial to the innovation process as a whole. Yet this process is often situated at the level of the business firm. Within the firm, organizational innovation is meant to improve the functional efficiency of the work processes and their use of technology. Economic innovation then is quite vague, referring to any innovation that will contribute to economic efficiency, including social innovation meant to improve the human relationships between agents working in the firm.

In sum, in the analysis of socio-technical systems and their innovation systems, the relationship between institutional change, technological innovation, economic growth and development is still predominantly considered as circular-causal, putting technological innovation at the nexus of development dynamics. In the sequel of the article, I seek to deconstruct these circular dynamics. I want to investigate what this deconstruction means for the place of diverse types of innovation and other human agencies in socio-economic development and what the consequences are for scientific research. I start below by briefly looking at diverse and evolving meanings of innovation as a concept and a human practice from the seventeenth century onwards.

THE HISTORY OF INNOVATION THOUGHT AND PRACTICE

Referring to Godin (2015) but also to my work with Bob Jessop, Lars Hulgård and Abdel Hamdouch (Jessop et al., 2013), I argue that innovation analysis when it started flourishing among academics say around 1950, and later on among policymakers, had lost touch with the history of thought and practice of innovation. In fact, the original meanings of innovation were much more social, political and cultural, slowly hosting technology in a social cultural womb as described in many powerful syntheses of the diverse worlds of development (Stöhr and Tödtling, 1978). But unfortunately, the technological baby rejected the womb and even imposed its transformation into a socio-technical system instead of a complex social (political) system.

The term innovation was already used in political and ideological debates back in seventeenth century England. Later on, innovation and, as of the early nineteenth century, social innovation began to refer to different forms of social change, or as some observers said: revolution. The differences in meaning of (social) innovation was also geopolitically marked. Whereas in England social innovation tended to be interpreted as revolution or a social political transformation threatening the social political order, in France it received a more positive interpretation as a driver of the necessary change of the social conditions of workers amidst the hardship of the industrial revolution.

By the beginning of the twentieth century social innovation came with a number of complementary meanings, all referring to various dimensions of what in contemporary development language would be called human development: satisfaction of basic needs, creation of associations to protect the rights of the workers and other interest groups, education for everybody, provision of public health services, and so on. Innovation, hence, was considered as any force or agency leading to the improvement of the human condition.

But as modernism became the ideology of the leading social forces in society, science and technology became main drivers of human progress. Generally speaking, as of the middle of the nineteenth century, gentrification of derelict quarters, installation of sewage systems and piped water provision, establishment of educational systems, legal and institutional frameworks entitling rights to groups who in very diverse ways were excluded from societal benefits, … were all aspects of a systemic engineering approach to the solution of the problems of humanity (Friedmann, 1987). Large-scale, top-down organizations, many of which were the outcome of workers' struggle, counting on the benefits of universality and large-scale inclusive systems became the norm for "modernizing" society and establishing its "social welfare system". In such hybrid large-scale approach to the organization of humanity, in the western world the state and corporate business found each other in an opportunistic marriage; in the communist world, top-down organization was filled in a different way, through central planning and the establishment of large scale production and distribution units.

In this invasive positive science-based approach to the organization of society, technology gained in prominence. This gain came with a shift in the meaning of innovation. By the 1950s, innovation was mainly understood as technological innovation. Innovation economics played a very important role in conceptualizing the nature of innovation and its connection to growth and development. The first models of growth and innovation in the neoclassical economic tradition did not analyze the nature of innovation at all: they simply established a link between innovation as a source of productivity growth and economic growth as produced by innovative firms. This first generation of analysis of the relationship between innovation and growth did not even refer to Schumpeter's work on the innovative firm; only an incrementalist reading of technology and growth, moving away from development concerns, mattered (Nelson, 1987).

Fortunately, economics is not a monolithic science, there are many streams; and although neoclassical and today neoliberal economics constitute the pillars of contemporary hegemonic economic discourse, other streams in economic thought and practice have had significant influence.

Two important developments within economic science have occupied a significant role in the reconsideration of the meaning of innovation in the development of society and its communities. These developments are still going on and seem to be converging and diverging at the same time. The first main development was the return of institutional and evolutionary economics at say the end of the 1980s. Institutional economics is older than neoclassical economics. It goes back to different streams in social science to say the middle of the nineteenth century. The German Historical school, the Scandinavian school, old American institutionalism, new institutionalism, ... all of them have examined the role of institutions in innovation and development (Moulaert and Sekia, 2003). Unfortunately, the most influential strands in institutional and evolutionary economics still consider institutions as instrumental to innovation and defined innovation primarily as technological. Yet, bit by bit, and especially in the spatial approaches to innovation, other dimensions or types of innovation have become more relevant: economic innovation – with an opening also to social economy, organizational innovation and social innovation – and (social) innovation in governance. My reading of recent literature on spatialized innovation systems is that, thanks to the attention given to territoriality, more attention is given to the cultural and social features of innovation processes and strategies, and that the instrumental relationships between institutions, innovation and development are no longer axiomatically posited. Indeed *territoriality* shows the social complexity, the cultural diversity and political economic conflictuality of the regions and localities in which innovation and development take place. Regions and territories have been in the focus of development theory and practice for long. The confrontation between development from above (top-down) and development from below approaches was an important topic of scientific discussion back in the 1970s. Walter Stöhr, for example, made a fine comparison between both approaches in which he stressed the role of ideology, culture, ethics and social relations in development; and explained clearly how modernist development approaches from above often alienated local social relations that were at the core of the bottom-up development analysis (Stöhr and Tödtling, 1978). In their seminal paper, Walter Stöhr and Franz Tödtling also provide a summary of various theories of uneven development and unequal exchange.

Since the 1970s significant scientific progress has been made in the analysis of National and Regional Innovation Systems. The analysis of culture and institutions has been fine-tuned and the focus shifted from instrumental relationships between regulations, institutions, innovation and competitiveness to co-learning and learning regions. By the end of the 1980s, and not the least because of the greater visibility of the economic, social and political problems of regions and cities in decline, the 'development from below' literature received a new impetus. Faced with problems of declining neighbourhoods and localities for which the regional innovation systems recipes did not work, new approaches stressing the diversity of assets, the role of social relations, of empowerment and the countervailing of development constraining powers, ... received more attention and came to life (Moulaert and Nussbaumer, 2014; Moulaert, 2000; Smith et al., 2016; Perera, 2016). Sociological and regulationist criticism of Territorial Innovation Models made a plea to give greater attention to the diversity of social relations in local and regional development analysis, to make a multi-dimensional analysis of the nature of innovation processes and strategies, governance and regulation, culture and economy, triggers of socio-political transformation (Moulaert, 1992). It is in this ambiance that our joint work on Integrated Area Development for cities and smaller localities, social innovation for local and regional development, the Social Region, took root and began to have an influence on urban and local development policy (Urban, Local Agenda 21, Sozialstadt, …) but especially on grassroots and activist initiatives (Moulaert, 2000; MacCallum et al., 2009).

But unfortunately – and in this short contribution a bit simplified as a representation of reality – the two trajectories addressing the role of innovation in development roll on following their own recipes. Innovation Systems and Regional Innovation Systems keeps heralding the *transmission logic from invention cum institutions to technological and economic innovations*, with new activities and improved competitiveness fostering growth and development. And socially innovative approaches to development stress the core role of social relations, the diversity of activities in a development agenda, the involvement of the local communities, the necessity of socio-political transformation to make local development happen, … The miscommunication or respectful coexistence between both trajectories has many explanations: path dependency of scientific approaches (more orthodox vs. more heterodox/pluralist economics with the latter showing more openness to interdisciplinarity), ideological continuity, lack of interest of the consultancy poisoned political world in changing policies from technological innovation to diversified development agendas, preference of financial capital to invest in high technology, professional services and up-market real estate development.

RECONNECTING INNOVATION WITH ITS SOCIAL ROOTS

As scientists, not policymakers, we may collaborate with or give advice to policymakers and activists; or combine roles with them. Yet our role is primarily scientific and pedagogical. Our purpose is to contribute to a comprehensive understanding of the factors and processes which foster human development in a territorial setting. This role should not be underestimated in its importance. We know from our previous work that discursive, analytical and practical uses and meanings of innovation have changed over the centuries. Such changes are also possible today. In the past, various political, social, cultural and scientific forces have played a role in the understanding of what innovation means. These dynamics are a mirror of the complexity of the changing society in which innovation has been aspired, pursued or has taken place. We ran into religious, political, customary, cultural, social economic and technological innovations. And even if the same terms were used in different periods, the concrete content varied. As much as it is important to understand the history of the social forces that have determined the different meanings and practices of social innovation, it is important today to understand the social forces that determine the various meanings of innovation today and how these forces could be influential in the building of a more development feasible (social) innovation concept. The history of the (social) innovation concept, we believe, can be used as a lead to reinterpret the different dimensions of innovation for human development in a territorial context. But to do such rebuilding of the innovation concept and its role in development, we will need to rely on different contributions from social science. The lower part of the figure below gives an overview of the history of the key concepts of social innovation and how these are related to the visions of societal development and the ideologies inspiring and reflecting them. The top part shows how, in a kind of playback mechanism, the historical categories of (social) innovation can be used to analyze contemporary meanings of (innovation) and its meaning for human development and socio-political transformation.

Geogarden harvest, Leuven © Constanza Parra

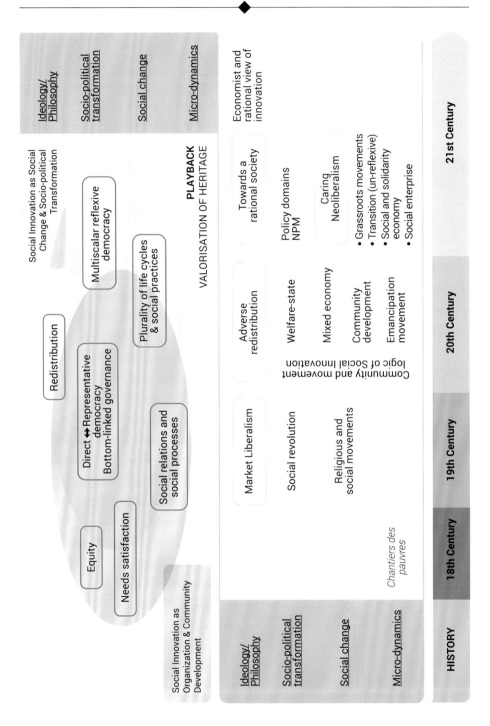

Source: Moulaert et al., 2017: 48

CHALLENGES FOR SOCIAL SCIENCE

In a way, the dominance of mainstream economics and the euphoria of increasingly new technological innovation have prolonged the rationalist and top-down culture inherent to modernity. Notwithstanding the positive heritage of modernity and Fordism for especially Western society (scientific progress to improve the general health condition of the population, productivity growth, and the welfare system), the prolonged dominance of modernity has left little opportunity to give more visibility to the role of diversity in human agency, the criticism of rational behaviour and rationality as a principle for governance, the necessity of socio-political transformation to countervail extreme power relations and to give a more effective place to bottom-up democracy, and so on. The economies of scale and scope which come with modernism and the capitalist economic system have been shown to be extremely instrumental to capitalist accumulation and disproportionate personal enrichment. These "power economics" certainly fenced off the access to post-modern critique and alternatives to top-down, instrumentalist approaches to development, as already criticized intelligently by Walter Stöhr.

Let us return to the main question: *How can social science re-enact the social character of innovation and development? How can it contribute to rebuild the place of (social) innovation in the analysis of societal development, at interconnected spatial scales?*

Several tasks, ongoing or ahead, are relevant:

- Reconnect collective human agency to development. The nineteenth century Historical School and also the twentieth century development theories offer significant material to do so. Recent new syntheses in social science, such as the cultural and institutional turn, have shown how collective and individual agency are institutionally and culturally conditioned, yet also develop autonomously according to new ideas, challenges, … One of the insights here is that macro-economic and macro-social (policy) analysis can no longer be analyzed in terms of broad categories of human collective agency, such as *innovation*, *regulation*, *social policy* but that other such categories should be considered with regard to their role and impact.

- Therefore, it is essential to situate innovation and policymaking among other categories of collective human agency fostering development, such as cultural emancipation, bottom-linked governance, participatory governance, education, solidarity movements. This could, among others, be done by further detailing the different meanings of social innovation as labelled in the figure. Social innovation indeed embraces a diversity of cultural, social and political features; it refers to micro-logic agency but also to collective human agency in various forms of emancipation, such as collective learning, pursuit of gender diversity and respect, building of solidarity forms of co-working, bottom-up governance practices, interactive practices fostering the nature-culture nexus, and so on. Recognizing and analyzing diversity in collective human behaviour also involves the identification and specification of rational and irrational types of behaviour, the way they materialize their intentions or desires, and their impact.

- Recognition of the diversification of human (collective) behaviour will also contribute to deconstruct *innovation idealism* showing so-called innovation processes as leading to denovation, trial and error, failure, undo innovation, return to *old* practices and technologies (Godin and Vinck, 2017).

- Reconsider the different meanings of institutions, governance, ... Starting from a broader discussion of the different characters of social relations, the relations between development, social change and innovation along other types of collective agency, institutions can be any form of instituted or established social relations. Informal institutionalization is not a contradiction, but a deep reality in society and economy. People spontaneously and intentionally develop codes of behaviour, habits, informal rules of interaction, ... which very often exist instead of formal (corporate, state) regulations, supplement or just overrule them. Analyzing these forms of informal institutionalization will contribute to the understanding of the pluralist economy as it really exists (Mingione, 1991).

- Research and pedagogical methodologies should be adapted to these more social and humane views of development and innovation. This implies more bottom-up models of education, stressing the role of ethics of cooperation and reciprocity. As to R&D, the research action model should become more organized along the lines of social innovation practice (see Moulaert and MacCallum, 2019). R&D itself should give more space to features of human development and social innovation in all spheres of society and economy.

CONCLUSION

Musing about the different historical meanings of innovation and social innovation is helpful in *deconstructing the nature of innovation systems and socio-technical systems*. There is a great need for innovation scholars to read and interpret the critical literature on (innovation) systems and their history. Trying to understand human development as the outcome of rational decision-making and processes leads to a misapprehension of how development really takes place. Real development consists of a multitude of processes and agencies pursuing diverse human ambitions converging towards greater economic, social and political cohesion or, more often so, diverging to greater inequality in diversity.

The following leads for future social science research should in my opinion be considered consciously:

- Innovation occurs in very diverse societies, each of which harbour a wide diversity in agency, social relations and institutions. This diversity cannot be grasped in terms of socio-technical and institutional trajectories only, it needs the analysis of the quintessence of social and cultural relationships, of the drivers of human behaviour, of norms and ethics, of the many faces of (structural) power relations, …

- Innovation has many faces. As the figure shows, its history is a pedagogical mirror to interrogate its different dimensions and its relations to other collective agencies, such as emancipation movements, socio-cultural association, governance and (other) co-production practices.

- Governance encompasses a social reality that cannot be grasped within the straightjacket of a rationalized systemic road-map. It is a complex hybrid of different modes of coordination, among which many are informal and emerging from spontaneous association and reciprocal social control.

- Future social science requires open-minded learning and research processes respecting the ethics of social innovation (respectful communication, association, reciprocity, …). Such ethics should also inspire the governance of educational systems.

- Ethical political leadership is essential in open-minded learning processes and the building of democratic bottom-linked governance.

- The analysis of and dealing with complexity of innovation and development requires interdisciplinary research. Social science cannot do it by itself, cooperation between various disciplines is needed. Recent synergies between social science, geography, planning, biology, ecology, …on ecological, biological, governance and community building issues are promising as breeding grounds for more systematic cooperation. ◀

Geogarden edible flowers, Leuven © Constanza Parra

42. A FRANK STORY ABOUT A CONVERSATION THAT (N)EVER TOOK PLACE

Han Verschure

(Being an engaged social innovation researcher aiming for political transformation, requires combining multiple roles, developing various profiles, coping with changing research and action contexts, dealing with increasing productivity demands and the pressure to publish, and handling the banalities of human relations. As such, the editors believe that the non-academic virtual dialogue on the merits and pitfalls of an academic career deserves a spot in a 'collection of research and analyses independent of the conventional academic norms and constraints'.)

H: For a while you identified yourself as a Flemish Primitive; that sounds both promising and creative, but maybe you have other meanings implied.

F: Well, when I returned from my long academic voyages abroad, I returned from an overseas New Castle to my home country and found myself as if I had been thrown back to the late middle-ages, you know with a real old castle and a mill, with Masters, knights, squires, servants, stable-boys, and a whole hierarchy of Castle Kings and Queens, Lords and Ladies-in-waiting, Bailiffs, even an occasional visiting Archbishop..... Knights, even though of good standing such as myself, even with the proper attire of a helmet which barely kept my wild curls in tow and with open visor (perhaps too open in a world of cleverly concealed ambitions), and even with the harness and all, we were only allowed to speak, let alone work together, with other knights if characters matched and were approved by the Masters. Otherwise you kept to your own turf, your visor closed, managed your own horses, got your own hay-stack, and so on.

H: But you underestimate yourself. The Flemish Primitives, they were actually at the avant-garde of artistic endeavours at the time; one of their real centres of excellence was Bruges, a town that also had the privilege to initiate your creativity.

They were innovators, masters of iconography; they liked complex pieces such as diptychs, triptychs or polyptychs. They inspired and exchanged with Italian artists. I agree, their work was often commissioned by the Church and the wealthy, but it was popular even among the commoners and the ordinary folks.

F: If you say so, but …

H: Would you have preferred to have adopted the name Don Quixote, also a famous historical figure, an avant-gardist, a freedom-fighter, a rebel-for-a good-cause, fighting injustice. Admittedly he was a bit mad, but aren't all geniuses a bit mad? And anyway, he was a harmless idealist as he was kept in balance by his follower Sancho Panza, who from time to time explained to him that all windmills do not look like giant enemies, that all potatoes do not look like cannonballs, that ordinary pubs may be as good places as lofty castles, that all church-addicts are not crooks, that soon his skinny horse would be replaced by a steel one which rides more easily … Of course, he also wanted to convince his Dulcinea that he was fearless. She was quite joyous and with sound laughter and magic hands to relieve peoples' pains. Sadly, the story goes they did not meet often enough, but I am sure she secretly loved him because of his madness. And let us not forget, he did liberate innocent people working in the fields from the hands of the unscrupulous Kings and Rulers. Supposedly their only *crime* was to plant the crops they knew would be healthy and not harm the common people.

F: Well, if you say so, but …

H: I know you will not easily give in to all suggestions coming along, so allow me to suggest another name to adopt. What about Robinson Crusoe? He stranded on an island after having survived several shipwrecks. He made himself a reasonable life against all odds in a seemingly hostile environment, getting creative with whatever nature had to offer. He started to like the lonely place. He occasionally saw real primitives coming to his island to perform offerings of humans, as a ritual. Well he had seen worse things in life so he got used to it, and besides that, meeting the locals was part of his way of working. His only problem was that he met a local whom he named Friday, after the only day of the week he always got nervous and lonely, and longed to return to his civilized home-world. After years, he made it back to his world, taking Friday along, but never knowing whether Friday would really be happy in such a civilized world. His adventures must certainly appeal to you.

F: Well, let me think, but ….

H: Ok, here is another proposal: Why not adopt the name Captain Ahab, the famous fellow who kept chasing the giant white whale Moby Dick. The Captain did have a good crew on board, although a quite cultural mixture of fellows and, exceptionally for a whaler, also some ladies; they were a good bunch. Once, several years ago, the white whale, almost caught, bit off part of the Captain's leg before disappearing into the deep. Ahab, now with one wooden leg, vowed never to give up, and kept chasing the whale around the world. His crew was very loyal, had a good time, cracked jokes, learned the tricks of the trade, learned to tie and untie the most complex knots, but at times they thought the Captain was getting a bit paranoid. Well unfortunately the story goes that with the final catch the white whale took the whole ship down into the depth of the seas.

F: No, that is too dramatic and not an honourable ending.

H: Oh, I see, you want a more honourable name-say. Well what about A Constant Gardener. That, I suggest could be quite appropriate, as it tells the story of a British Diplomat named Quayle, a very honourable man, who likes to tend his garden meticulously, but becomes a lone truth-seeker when he finds out about conspiracies of multinationals to test drugs which produce quite severe, even deadly, side effects; They test these drugs on innocent Africans whose life is seen by the pharmaceutical industries as "interesting experiment material". Sitting quietly in his attic room full of books, he is compelled to go into the real world to seek the truth. He sadly finds out that crooks can be more innovative than honest people. He is all the more dismayed as he finds out that even the British Foreign Office is part of the conspiracy. The author of the book, my favoured John Le Carré, says himself about his story: 'by comparison with the reality, my story [is] as tame as a holiday postcard'.

At the end of Quayle's career, I am sure, if he had survived, he would have remained a constant gardener, tending his vegetables, his herbs, battling the snails, occasionally bending his back too long so that his lovely wife could treat him to 'the medium is the massage'. And by the way, John Le Carré is also a good cook and has written a recipe booklet. He can teach you how to make the béarnaise catch on (béarnaise moet pakken).

F: Well that sounds like an appealing career end, but frankly speaking, would you mind if I continue from time to time to stop by the castle and the mill and to still act as all the above suggested famous people? You know I like the human complexity and particularly this mixture of scientific facts and - at times cynical - fiction.

H: Quite so, I am sure your crew will love to see you regularly coming, if you occasionally organize a party. Of course, you may find yourself, after some years at a loss, looking for the castle. By then it will have been torn down to make way for a twenty storey building which some of our colleagues would love to design. The park around will then be a giant parking lot, even the city will likely be renamed Louisville, by the local folks nicknamed (i)MacLouis.

However, knowing you, you will continue to explore the wide world, so, send us a holiday card from time to time, whether from your hometown in the far-west, from a lovely Greek island, from Turkish ruins, from a reputed Italian city, from the cold plains of Manitoba, or from the top of the Cotopaxi which, I am sure, you want to climb once again. And, please, do not forget to send us a message from this peculiar island – if you ever find it – called Utopia. ◄

APPENDIX

LIST OF AUTHORS

Louis Albrechts, KU Leuven, Belgium

Isabel André, CEG, Universidade de Lisboa, Portugal

Seppe De Blust, KU Leuven, Belgium

Iratxe Calvo Mendieta, Laboratoire Territoires, Villes, Environnement, Société, France

Stuart Cameron, Newcastle University, UK

Lucia Cavola, ITER srl, Italy

Daniela Coimbra de Souza, Secretaria Municipal de Transportes de São Paulo, Brazil

Giancarlo Cotella, Politecnico di Torino, Italy

Pavlos-Marinos Delladetsimas, Harokopio University of Athens, Greece

Michael Edwards, University College London, UK

Brenda Galvan-Lopez, Freiburg Planning Department, Germany

Marisol García, Universitat de Barcelona, Spain

Hubert Gulinck, KU Leuven, Belgium

Patsy Healey, Newcastle University, UK

Jean Hillier, RMIT University, Australia

Felicitas Hillmann, Leibniz Institute for Research on Society and Space, Germany

Bob Jessop, Lancaster University, UK

Michael Kaethler, KU Leuven, Belgium

Giota Karametou, Harokopio University of Athens, Greece

Chris Kesteloot, KU Leuven, Belgium

Ahmed Z. Khan, Université Libre de Bruxelles, Belgium

Juan-Luis Klein, CRISES, Université du Québec à Montréal, Canada

Annette Kuhk, KU Leuven, Belgium

Maarten Loopmans, KU Leuven, Belgium

Diana MacCallum, Curtin University, Australia

Margaret Macharia, Technical University of Kenya, Kenya

Albert Martens, KU Leuven, Belgium

Flavia Martinelli, Università degli Studi Mediterranea di Reggio Calabria, Italy

Abid Mehmood, Cardiff University, UK

Konrad Miciukiewicz, University College London, UK

Emmanuel Midheme, Maseno University, Kenya

Kevin Morgan, Cardiff University, UK

Enrica Morlicchio, Università degli Studi di Napoli Federico II, Italy

Frank Moulaert, KU Leuven, Belgium

Andreas Novy, Wirtschaftsuniversität Wien, Austria

Stijn Oosterlynck, Universiteit Antwerpen, Belgium

Angeliki Paidakaki, KU Leuven, Belgium

Constanza Parra, KU Leuven, Belgium

Marc Pradel, Universitat de Barcelona, Spain

Jonathan Pratschke, Università degli Studi di Salerno, Italy

Prathiwi Widyatmi Putri, Copenhagen University, Denmark

Patricia Rego, Universidade de Évora, Portugal
Abdur Rehman Cheema, Rural Support Programmes Network, Pakistan
Carlos Rodrigues, Universidade de Aveiro, Portugal
Artur Da Rosa Pires, Universidade de Aveiro, Portugal
Jan Schreurs, KU Leuven, Belgium
Ruth Segers, KU Leuven, Belgium
Loris Servillo, KU Leuven, Belgium and University College London, UK
Ngai-Ling Sum, Lancaster University, UK
Erik Swyngedouw, University of Manchester, UK
Chiara Tornaghi, Coventry University, UK
Pieter Van den Broeck, KU Leuven, Belgium
Barbara Van Dyck, Université Libre de Bruxelles, Belgium
Han Verschure, KU Leuven, Belgium
Thomas Werquin, Axe Culture, Think Tank de Lille et de sa région, France

BIBLIOGRAPHY

Adorno, T. W. (1997), Aesthetic Theory, A&C Black.

Albrechts, L. (2011), 'Transformative practices: Where strategic spatial planning meets social innovation', in S. Oosterlynck, J. Van den Broeck, L. Albrechts, F. Moulaert and A. Verhetsel (eds), Strategic Spatial Projects: Catalysts for Change, London: Routledge, RTPI Library Series, pp. 17-25.

Albrechts, L. (2015), 'Ingredients for a more radical strategic spatial planning', Environment and Planning B: Planning and Design, 42 (3), 510-525.

Albrechts, L. (2017), 'Some ingredients for revisiting strategic spatial planning', in L. Albrechts, A. Balducci and J. Hillier (eds), Situated Practices of Strategic Planning. An international perspective. Abingdon/New York: Routledge, pp. 389-404.

Albrechts, L., A. Balducci and J. Hillier (eds) (2017), Situated Practices of Strategic Planning, London: Routledge.

Álvarez, L. (2013), 'Indignados en el cine: 20 películas que reflejan el movimiento del 15-M', accessed 15 May 2013 at http://extracine.com/2013/05/indignados-en-el-cine.

André, I. and A. Abreu (2009), 'Social creativity and post-rural places: the case of Montemor-o-Novo, Portugal', Canadian Journal of Regional Science/Revue Canadienne des Sciences Régionales, 32 (1), 101-114.

André, I., E. B. Henriques and J. Malheiros (2009), 'Inclusive places, arts and socially creative milieux', in D. MacCallum, F. Moulaert, J. Hillier and S. Vicari-Haddock (eds), Social Innovation and Territorial Development, Aldershot: Ashgate Publishing, pp. 149-166.

André, I., J. Malheiros and A. Carmo (2013), 'The rythm of arts in the socially creative city', in J.L. Klein and M. Roy (eds), Pour une nouvelle mondialisation: le défi d'innover, Montreal: Presses de l'Université du Québec, pp. 191-207.

André, I., P. Rêgo, S. Pedro-Rêgo and B. Osório (2014), Inovação Social no Terceiro Setor - O Distrito de Évora, Évora: Fundação Eugénio de Almeida.

Arocena, R. and J. Sutz (2003), 'Inequality and innovation as seen from the South', Technology in Society, 25 (2), 171-182.

Aubouin, N. and E. Coblence (2013), 'Les nouveaux territoires de l'art, entre îlot et essaim. Piloter la rencontre entre friche artistique et territoire', Territoire en mouvement Revue de géographie et aménagement. Territory in movement Journal of Geography and Planning, 1 (17-18), 91-102.

Australian Government, Department of Employment (2013), Social Innovation: Introduction to social innovation, accessed 10 January 2014 at https://docs.employment.gov.au/system/files/doc/other/socialinnovtnintronov2013_2.pdf.

Bailey, N. (2012), 'The role, organization and contribution of community enterprise to urban regeneration policy in the UK', Progress in Planning, 77 (1), 1-35.

Benneworth, P. and J. Cunha (2015), 'Universities' contributions to social innovation: reflections in theory & practice', European Journal of Innovation Management, 18 (4), 508-527.

Bentham, J., A. Bowman, M. de la Cuesta, E. Engelen, I. Ertürk, P. Folkman, J. Froud, S. Johal, J. Law, A. Leaver, M. Moran and K. Williams (2013), 'Manifesto for the Foundational Economy', CRESC Working Paper 131, University of Manchester.

Bertolini, L. (2010), 'Coping with the irreducible uncertainties of planning: An evolutionary approach', in J. Hillier and P. Healey (eds), The Ashgate Research Companion to Planning Theory: Conceptual Challenges for Spatial Planning, Farnham (Surrey, England): Ashgate Publishing Limited, pp. 413-424.

Block, F. (2001), 'Introduction', in K. Polanyi, The Great Transformation: The Political and Economic Origins of Our Time, Boston: Beacon Press, pp. xviii-xxxviii (Originally published: New York: Farrar and Rinehart, 1944).

Boichot, C. (2013), 'Les espaces de la création artistique à Paris et Berlin. Entre pôle artistique et centralité urbaine', Territoire en mouvement Revue de géographie et aménagement. Territory in Movement Journal of Geography and Planning, (19-20), 19-39.

Bollen, R. and F. Moulaert (1983), Racisten hebben Ongelijk, Leuven: Kritak.

Boltanski, L. and L. Thévenot (1991), De la justification. Les économies de la grandeur, NRF essais, Gallimard.

Borja, J. (2011), Ciudades del mañana. Derecho a la ciudad y democracia real, Cafe de las ciudades. Barcelona: IDHC.

Bouchard, M. J., (ed.) (2013), Innovation and the Social Economy, Toronto: University of Toronto Press.

Brans, M., D. Jacobs, M. Martiniello, A. Rea, M. Swyngedouw, I. Adam, P. Balancier, E. Florence and T. Van der Straeten (eds) (2004), Recherche et politiques publiques: le cas de l'immigration en Belgique. Onderzoek en beleid: de gevalstudie van immigratie in België ('Research and Policy: Case study of Immigration in Belgium'), Ghent: Academia Press.

Brown, V. and J. Lambert (2013), Collective Learning for Transformational Change: A Guide to Collaborative Action, New York: Routledge.

Bureau of European Policy Advisors (BEPA) (2010), Empowering people, driving change: Social Innovation in the European Union, Brussels: EC.

Bureau of European Policy Advisors (BEPA) (2011), Empowering people, driving change: Social innovation in the European Union, Luxembourg: Publications Office of the European Union.

Calvo-Mendieta, I. (2005), L'économie des ressources en eau: de l'internalisation des externalités à la gestion intégrée. L'exemple du bassin versant de l'Audomarois (Doctoral dissertation, Université des Sciences et Technologie de Lille-Lille I)

Cameron, S. and S. González (2013), 'Newcastle-upon-Tyne and the Northern Way; neo-liberal responses to uneven development in the North of England', in F. Martinelli, F. Moulaert and A. Novy (eds), Urban and Regional Development Trajectories in Contemporary Capitalism, Abingdon: Routledge, pp. 264-283.

Capel, H. (1996), 'La rehabilitación y el uso del patrimonio histórico industrial', Documents d'analisi geografica, (29), 19-50.

Cassinari, D., J. Hillier, K. Miciukiewicz, A. Novy, S. Habersack, D. MacCallum and F. Moulaert, (2011). 'Transdisciplinary Research in Social Polis'. Social Polis Working

Paper.

Cassinari, D. and F. Moulaert (2014), 'Enabling transdisciplinary research on social cohesion in the city: The Social Polis experience', in E. A. Silva, P. Healey, N. Harris and P. Van den Broeck (eds), The Routledge Handbook of Planning Research Methods, London: Routledge, pp. 414-425.

Chesbrough, H. (2006), Open Business Models: How to Thrive in the New Innovation Landscape, Boston: Harvard Business Press.

Christiaens, E., F. Moulaert, F. and B. Bosmans (2007), 'The end of social innovation in urban development strategies? The case of Antwerp and the neighbourhood association "BOM"', European Urban and Regional Studies, 14 (3), 238-251.

Cooke, P., F. Moulaert, E. Swyngedouw, O. Weinstein and P. Wells (1992), Towards Global Localisation: The Computing and Telecommunications Industries in Britain and France, London: UCL Press.

Cools, P. and S. Oosterlynck (2015), 'Case Study 13: Energy poverty and social entrepreneurship: strengthening pre-financing models for energy efficient electrical appliances through the 'Energy for All' programme', accessed 1 August 2015 at https://repository.uantwerpen.be/docman/irua/c759a3/130573.pdf

Cornwall, A. (2008), 'Unpacking "Participation": models, meanings and practices', Community Development Journal, 43 (3), 269-283.

Crouch, C. (2015), The Knowledge Corrupters. Hidden Consequences of the Financial Takeover of Public Life, Cambridge: Polity.

Dale, G. (2016), Reconstructing Karl Polanyi, London: Pluto Press.

Danermark, B., M. Ekström, L. Jakobsen and J. C. Karlsson (2005), Explaining Society. Critical realism in the social sciences, London/New York: Routledge.

Davoudi, S. and A. Madanipour (eds) (2015), Reconsidering Localism, London: Routledge.

Dawney, L., S. Kirwan and J. Brigstocke (eds) (2016), Space, Power and the Commons: The struggle for alternative futures, London: Routledge.

Dewaelheyns, V., K. Bomans and H. Gulinck (eds) (2011), The Powerful Garden. Emerging views on the garden complex, Antwerp: Garant Publishers.

Dunsire, A. (1996), 'Tipping the balance: autopoiesis and governance', Administration & Society, 28 (3), 299-334.

Edwards, M. (2016), 'The Housing crisis and London', City, 20 (2), 222-237.

Eizaguirre, S., M. Pradel, A. Terrones, X. Martinez-Celorrio and M. García (2012), 'Multilevel governance and social cohesion: Bringing back conflict in citizenship practices', Urban Studies, 49 (9), 1999-2016.

Elliott, G. (2013), 'Character and impact of social innovation in higher education', International Journal of Continuing Education and Lifelong Learning, 5 (2), 71.

Featherstone, D., A. Ince, D. Mackinnan, K. Strauss and A. Cumbers (2012), 'Progressive localism and the construction of political alternatives', Transactions of the Institute of British Geographers, 37 (2), 177-182.

Florida, R. (2002), The Rise of the Creative Class—and How it is Transforming Leisure, Community and Everyday Life, New York: Basic Books.

Fontan, J. M. (2011), 'Innovation et transformation des sociétés: rôle et fonction de l'innovation sociale', Économie et Solidarités, 41 (1-2), 9-27.

Fontan, J. M., J. L. Klein and D. G. Tremblay (1999), Entre la métropolisation et le village global: les scènes territoriales de la reconversion. Québec: Presses Universitaires du Québec.

Fontan, J. M., J. L. Klein and D. G. Tremblay (2005), Innovation socioterritoriale et reconversion économique: le cas de Montréal, Paris: Editions L'Harmattan, Volume 34.

Friedmann, J. (1987), Planning in the Public Domain: From knowledge to action, Princeton University Press.

Friedmann, J. (2011), Insurgencies: Essays in Planning Theory, London: Routledge.

Garcia, M. (2006), 'Citizenship practices and urban governance in European cities', Urban Studies, 43 (4), 745-765.

Gibson-Graham, J. K. and G. Roelvinck (2009), 'Social innovation for Community Economies', in D. MacCallum, F. Moulaert, J. Hillier and S. Vicari-Haddock (eds), Social Innovation and Territorial Development, Aldershot: Ashgate Publishing.

Godin, B. (2015), Innovation Contested: The idea of innovation over the centuries, Routledge.

Godin, B. and D. Vinck (eds) (2017), Critical Studies of Innovation: Alternative approaches to the Pro-Innovation Bias, Cheltenham: Edward Elgar Publishing.

González, S., F. Martinelli and F. Moulaert (2010), 'ALMOLIN: How to analyse social innovation at the local level?', in F. Moulaert, F. Martinelli, E. Swyngedouw and S. González (eds), Can Neighbourhoods Save the City? Community Development and Social Innovation, London: Routledge, pp. 49-67.

Gramsci, A. (1971), Selections from the Prison Notebooks, London: Lawrence & Wishart.

Gualini, E. (2001), Planning and the Intelligence of Institutions. Interactive approaches to territorial policy-making between institutional design and institution-building, Aldershot: Ashgate Publishing.

Hamdouch, A. and F. Moulaert (2006), 'Knowledge infrastructure, innovation dynamics, and knowledge creation/diffusion/accumulation processes: a comparative institutional perspective', Innovation: The European Journal of Social Science Research, 19 (1), 25-50.

Healey, P. (1997), Collaborative Planning: Shaping places in fragmented societies, London: MacMillan.

Healey, P. (1999), Institutionalist analysis, communicative planning and shaping places, Journal of Planning Education and Research, 19 (2), 111-122.

Healey, P. (2009a), 'In search of the "strategic" in spatial strategy making', Planning Theory and Practice, 10 (4), 439-457.

Healey, P. (2009b), 'The pragmatic tradition in planning thought', Journal of Planning Education and Research, 28 (3), 277-292.

Healey, P. (2015a), 'Civil society enterprise and local development', Planning Theory & Practice, 16 (1), 11-27.

Healey, P. (2015b), 'Citizen-empowered local development initiative: recent English experience', International Journal of Urban Sciences, 19 (2), 109-118.

Healey, P. (2015c), 'Civic capacity, place governance and progressive localism', in S. Davoudi and A. Madanipour (eds), Reconsidering Localism, London: Routledge, pp. 105-125.

Healey, P. (2018), 'Creating public value through caring for place', Policy & Politics, 46 (1), 65-79.

Hillier, J. (2003), 'Agonizing over consensus: Why Habermasian ideals cannot be "real"', Planning Theory, 2 (1), 37-59.

Hillier, J. (2007), Stretching beyond the Horizon: A Multiplanar Theory of Spatial Planning and Governance, Aldershot: Ashgate Publishing.

Hillier, J. (2011), 'Strategic navigation across multiple planes: Towards a Deleuzean-inspired methodology for strategic spatial planning', Town Planning Review, 82 (5), 503-527.

Hobin, V. and F. Moulaert (1986), Witboek integratiebeleid inzake migranten in Vlaanderen-België ('Integration Policy White Paper concerning Migrants in Belgian Flanders'), Colloquium 28 November 1986 Final edit. Brussels

Houston, D., D. MacCallum, W. Steele and J. Byrne (2016), 'Climate cosmopolitics and the possibilities for urban planning', Nature and Culture, 11 (3), 259-277.

Hubeau, B., A. Martens and T. Wouters (1989), Integratiebeleid Buitenlandse Minderheden in Vlaanderen-België. Onderzoeksvoorstellen buitenlandse Minderheden ('Integration Policy on Foreign Minorities in Belgian Flanders'), Leuven: KU. Departement Sociologie & Antwerp: UIA Departement Rechten.

Interfederaal gelijkekansencentrum (2015), Socio-economische Monitoring. Arbeidsmarkt en origine (Interfederal Centre for Equal Opportunities Socio-economic Monitor. Labour Market and Origin), accessed 2016 at www.diversiteit.be; now at: https://www.unia.be/en.

Jenson, J. (2015), 'Social innovation: Redesigning the welfare diamond', in A. Nicholls, J. Simon and M. Gabriel (eds), New Frontiers in Social Innovation Research, Hampshire: Palgrave Macmillan, pp. 89-106.

Jessop, B. (2001), 'Regulationist and autopoieticist reflections on Polanyi's account of market economies and the market society', New Political Economy, 6 (2), 213-232.

Jessop, B. (2007), 'Knowledge as a fictitious commodity: insights and limits of a Polanyian analysis', in A. Buğra and K. Ağartan (eds), Reading Karl Polanyi for the 21st century. Market Economy as a Political Project, Basingstoke: Palgrave, pp. 115-134.

Jessop, B. (2011), 'Metagovernance', in A. Bevir (ed.), Handbook of Governance, London: SAGE, pp. 106-123.

Jessop, B. (2016), 'Territory, politics, governance and multispatial metagovernance', Territory, Politics, Governance, 4 (1), 8-32.

Jessop, B. and E. Swyngedouw (2005), 'Regulation-Reproduction-Governance', DEMOLOGOS Working Paper.

Jessop, B., F. Moulaert, L. Hulgard and A. Hamdouch (2013), 'Social innovation research: a new stage in innovation analysis?', in F. Moulaert, D. MacCallum, A. Mehmood and A. Hamdouch (eds), International Handbook of Social Innovation: Collective Action, Social Learning and Transdisciplinary Research, Cheltenham: Edward Elgar Publishing, pp. 110-130.

Jones, M. R. (1997), 'Spatial selectivity of the state? The regulationist enigma and local struggles over economic governance', Environment and Planning A, 29 (5), 831-864.

Kamensky, J. M., T.J. Burlin (eds) (2004), Collaboration using Networks and Partnerships, Rowman & Littlefield Publishers: Lanham, MD.

Kemmis, S. and R. McTaggart (2000), 'Participatory Action Research', in N. Denzin and Y. Lincoln (eds), Handbook of Qualitative Research, Thousand Oaks: Sage.

Kesteloot, C. (2013), 'Sociospatial fragmentations and governance', in E. Corijn, E. and J. Van de Ven (eds), The Brussels Reader; A small world city to become the capital of Europe, Brussels: VUB Press, pp. 110-149.

Khan, A. Z., F. Moulaert and J. Schreurs (2013), 'Epistemology of space: Exploring relational perspectives in planning, urbanism, and architecture', International Planning Studies, 18 (3-4), 287-303.

Khan, A. Z., F. Moulaert, J. Schreurs and K. Miciukiewicz (2014), 'Integrative spatial quality: A relational epistemology of space and transdisciplinarity in urban design and planning', Journal of Urban Design, 19 (4), 393-411.

Kimaro, O., M. Shilereyo, B. Mwamengo, H. Gulinck and G. Walalaze (2014), 'Communities' consideration underlying vegetation edge ecosystem in Magamba village, Lushoto District, Tanzania', Journal of Land and Society, 1 (1), 67-81.

Klein, J. L. and D. Harrisson (eds) (2006), L'innovation sociale: émergence et effets sur la transformation des sociétés. Québec: Presses de l'Université du Québec.

Klein, J. L., J. L. Laville and F. Moulaert (eds) (2014), L'innovation sociale, Toulouse: Érès.

Klein, N. (2015), Tout peut changer, Montréal: Lux Éditeur.

Kuhk, A., J. Schreurs and M. Dehaene (2015), 'Collective learning experiences in planning: The potential of experimental living labs', paper presented at the AESOP Conference, Prague, July 2015.

Kuhk, A., M. Dehaene, M. Dumont and J. Schreurs (2016), 'Toekomstverkenning als collectief leren. Onderzoek naar planning in het licht van onzekerheid en complexiteit', Rapport WP3. Brussel: Steunpunt Ruimte.

Kunnen, N., D. MacCallum and S. Young (2013), 'Research strategies for assets and strengths based community development', in F. Moulaert, D. MacCallum, A. Mehmood and A. Hamdouch (eds), The International Handbook on Social Innovation: Collective Action, Social Learning and Transdisciplinary Research, London: Edward Elgar Publishing, pp. 285-298.

Lambooy, J. G. (1988), 'Development trajectories of regions', in Z. Chojnicki (ed.), Concepts and Methods in Geography, Poznan: Adam Mickiewicz University, pp. 131-147.

Landry, C. (2000), The Creative City: A toolkit for urban innovators, London: Earthscan.

Le Floc'h, M. (2015), Plan-Guide. Arts et Aménagement des Territoires, France: POLAU, Pole des Arts Urbains, Ministère de la Culture et de la Communication.

Lévesque, B. (2014), 'Un monde qui se défait, un monde à reconstruire', in B. Lévesque, J. M. Fontan and J. L. Klein (eds), L' innovation sociale: les marches d'une construction théorique et pratique, Québec: Presses de l'Université du Québec, pp. 369-386.

Lévesque, B., J. M. Fontan and J. L. Klein (eds) (2014), L' innovation sociale: les marches d'une construction théorique et pratique, Québec: Presses de l'Université du Québec.

Ley, D. (2003), 'Artists, aestheticisation and the field of gentrification', Urban Studies, 40 (12), 2527-2544.

MacCallum, D., F. Moulaert, J. Hillier and S. Vicari (eds) (2009), Social Innovation and Territorial Development, Farnham: Ashgate Publishing.

Macguire, P. (1987), Doing Participatory Research: A Feminist Approach, Amherst: Centre for International Education, University of Massachusetts.

Macharia, M. and P. Van den Broeck (2016), 'Surviving through informality in Nairobi: the case of urban traders in Eastleigh's commercial centre', in R. Segers, P. Van den Broeck, A. Kahn, F. Moulaert, J. Schreurs, B. De Meulder, K. Miciukiewicz, G. Vigar and A. Madanipour (eds), The Spindus Handbook for Spatial Quality: A Relational Approach, Brussels: Academic and Scientific Publishers (ASP), pp. 212-221.

Manganelli, A. (2019), Unlocking Socio-Political Dynamics of Alternative Food Networks through a Hybrid Governance Approach. Highlights from the Brussels-Capital Region, and Toronto (Doctoral dissertation, KU Leuven)

Margolin, V. and R. Buchanan (1995), The Idea of Design. Cambridge, MA: The MIT Press.

Marston, G. (2004), Social Policy and Discourse Analysis: Policy change in public housing, Aldershot: Ashgate Publishing.

Martens, A. (1973), 25 jaar wegwerparbeiders. Het Belgische immigratiebeleid na 1945 ('25 years of disposable workers. Belgian immigration policy after 1945'), Leuven: KU Leuven Sociologisch onderzoeksinstituut. Rapport 1973/2 (doctoral thesis).

Martens, A. (2012), Onderzoek naar discriminatie in historisch perspectief ('Study of discrimination in an historical perspective'). Diversiteits-barometer Werk ('Employment Diversity Barometer'). Brussels: Centrum voor gelijkheid van kansen en racismebestrijding (Centre for Equal Opportunities and Opposition to Racism), pp.130-140, accessed 2016 at www.diversiteit.be ; now at https://www.unia.be/en.

Martens, A. and F. Moulaert (1978), Minder Vreemde meer Belgische arbeid? ('Less Foreign, more Belgian Labour?'), Nederlands Economisch Instituut, Rotterdam, Economische en Statistische berichten, 22-XI-1978, 63°jrg.,n° 3181 ('Economic and Statistical Bulletins' 22-XI-1978, vol. 63, no. 3181), pp. 1194-1196.

Martens A. and F. Moulaert (1981), Het immigratiebeleid van de Belgische overheid ('The immigration policy of the Belgian government'), in Vereniging voor Economie: Overheidsinterventies, Effectiviteit en Efficiëntie. ('Government Interventions: Effectiveness and Efficiency'), Fifteenth Flemish Economic Congress (Vlaamse Wetenschappelijk Economisch Congres), Leuven 8-9 May 1981, pp. 133-145.

Martens A. and F. Moulaert (eds) (1985), Buitenlandse Minderheden in Vlaanderen-België ('Foreign Minorities in Belgian Flanders'), De Nederlandsche Boekhandel, 17-30; 169-181.

Martinelli, F. (2008), 'Papa ERASMUS', in J. Van den Broeck, F. Moulaert and S. Oosterlynck (eds), Empowering the Planning Fields, Liber Amicorum Louis Albrechts, Leuven: Acco, pp. 223-226.

Martinelli, F. (2010), 'Historical roots of social change: philosophies and movements', in

F. Moulaert, F. Martinelli, E. Swyngedouw and S. González (eds), Can Neighbourhoods Save the City? Community Development and Social Innovation, Oxford/New York: Routledge, pp. 17-48.

Martinelli, F. (2012), 'Social innovation or social exclusion? Innovating social services in the context of a retrenching welfare state', in H. W. Franz, J. Hochgerner and J. Howaldt (eds), Challenge Social Innovation. Potentials for Business, Social Entrepreneurship, Welfare and Civil Society, Berlin: Springer, pp. 169-180.

Martinelli, F., F. Moulaert and S. González (2010), 'Creatively designing urban futures: a transversal analysis of socially innovative case studies', in F. Moulaert, F. Martinelli, E. Swyngedouw and S. González (eds), Can Neighbourhoods Save the City? Community Development and Social Innovation, Oxford/New York: Routledge, pp. 198-218.

Martinelli, F., F. Moulaert and A. Novy (2013), Urban and Regional Development Trajectories in Contemporary Capitalism, London: Routledge.

Martinelli, F., A. Anttonen and M. Mätzke (eds) (2017), Social Services Disrupted: Changes, challenges and policy implications for Europe in times of austerity, Edward Elgar Publishing.

Massey, D. (2005), For Space, Thousand Oaks CA: Sage.

Mehmood, A., and C. Parra (2013), 'Social innovation in an unsustainable world', in F. Moulaert, D. MacCallum, A. Mehmood and A. Hamdouch (eds), The International Handbook on Social Innovation, Cheltenham: Edward Elgar Publishing, pp. 53-66.

Meuleman, L. (2008), Public Management and the Metagovernance of Hierarchies, Networks and Markets, Heidelberg: Springer.

Miciukiewicz, K., F. Moulaert, A. Novy, S. Musterd and J. Hillier (2012), 'Introduction: problematising urban social cohesion: A transdisciplinary endeavour', Urban Studies, 49 (9), 1855-1872.

Midheme, E. and F. Moulaert (2013), 'Pushing back the frontiers of property: Community land trusts and low-income housing in urban Kenya', Land Use Policy, 35, 73-84.

Mingione, E. (1991), Fragmented Societies: A sociology of economic life beyond the market paradigm, Blackwell.

Montgomery, T. (2016), 'Are Social innovation paradigms incommensurable?', Voluntas, 27 (4), 1979-2000.

Moore, S. A. and A. Karvonen (2008), 'Sustainable architecture in context: STS and design thinking', Science Studies, 21 (1), 29-46.

Morgan, K. (1997), 'The learning region: institutions, innovation and regional renewal', Regional Studies, 31 (5), 491-503.

Morgan, K. (2015), 'The moral economy of food', Geoforum, 65, 294-296.

Moulaert, F. (1987), 'An institutional revisit to the Storper Walker theory of labour', International Journal of Urban and Regional Research, 11 (3), 309-330.

Moulaert, F. (1992), 'Services: The bridge between computing and communications', in P. Cooke, F. Moulaert, O. Weinstein, E. Swyngedouw and P. Wells (eds), Towards Global Localization: The Computing and Telecommunications Industries in Britain and France, London: UCL Press, pp 178-199.

Moulaert, F. (1996), 'Rediscovering spatial inequality in Europe: building blocks for an appropriate "regulationist" analytical framework', Environment and Planning D: Society and Space, 14 (2), 155-179.

Moulaert, F. (2000), Globalization and Integrated Area Development in European Cities, Oxford: Oxford University Press.

Moulaert, F. (2005), 'Institutional economics and planning theory: A partnership between ostriches?', Planning Theory, 4 (1), 21-32.

Moulaert, F. (2007), Social Innovation, Governance and Community Building: SINGOCOM Final Report, Luxembourg: Office for Official Publications of the European Communities.

Moulaert, F. (2009), 'Social innovation: Institutionally embedded, territorially (re) produced', in D. MacCallum, F. Moulaert, J. Hillier and S. Vicari (eds), Social Innovation and Territorial Development, Farnham: Ashgate Publishing, pp. 11-23.

Moulaert, F. (2010), 'Social innovation and community development: concepts, theories and challenges', in F. Moulaert, F. Martinelli, E. Swyngedouw and S. González (eds), Can Neighbourhoods Save the City? Community Development and Social Innovation, Oxford/New York: Routledge, pp. 4-16.

Moulaert, F. (2011), 'When solidarity boosts strategic planning', in S. Oosterlynck, J. Van den Broeck, L. Albrechts, F. Moulaert and A. Verhetsel (eds), Strategic Spatial Projects: Catalysts for change, London: Routledge, RTPI Library Series, pp. 79-84.

Moulaert, F. (2016), 'Radical social innovation', plenary lecture at WISERD CRESC Radical Social Innovation Colloquium. 19 May.

Moulaert, F. and A. Martens (1982), Arbeidsproces, sectorale dynamiek en gastarbeid in België (1970-1977) ('Labour process, sector dynamics and guest workers in Belgium (1970-1977)'). In J. M. Van Amersfoort and J. B. Entzinger, Immigrant en Samenleving ('Immigrant and Society'). Mens en Maatschappij, 57ste jrg. ('People and Society', vol. 57), Deventer: Van Loghum Slaterus, pp. 77-98.

Moulaert, F. and A. Martens (1985), De toestand van de buitenlandse minderheden in Vlaanderen en België. Een wetenschappelijke benadering ('The status of foreign minorities in Flanders and Belgium. A scientific approach'), in A. Martens and FF. Moulaert (eds), Buitenlandse Minderheden in Vlaanderen-België ('Foreign Minorities in Belgian Flanders'), De Nederlandsche Boekhandel, pp. 17-30.

Moulaert, F. and A. Martens (1986), 'Situation des minorités étrangères en Flandre et en Belgique' (The situation of foreign minorities in Flanders and Belgium), Tribune immigrée, July-October, (16-17), 17-22.

Moulaert, F. and A. Martens (1988), La main-d'oeuvre étrangère dans l'économie belge (1970-1981) a-t-elle accrue la flexibilité de l'emploi? (Has the foreign workforce in the Belgian economy (1970-1981) increased employment flexibility?) Une analyse par groupe professionnel (An analysis by occupational category), Colloque CNRS Mutations économiques et travailleurs immigrés dans les pays industriels, (CNRS colloquium, Economic transformations and immigrant workers in the industrial countries), Vaucraisson Paris, 28-30/1/1988 (27 p.)

Moulaert, F. and E. Swyngedouw (1989), 'A regulation approach to the geography of the flexible production system', Environment and Planning D: Space and Society, 7 (3), 327-345.

Moulaert, F. and P. W. Daniels (1991), The Changing Geography of Advanced Producer Services, London: Belhaven Press.

Moulaert, F. and E. Swyngedouw (1991), 'Regional development and the geography of the flexible production system. Theoretical arguments and empirical evidence', in U. Hilpert (ed.), Regional Innovation and Decentralization, London/New York: Routledge.

Moulaert, F. and E. Swyngedouw (1992), 'Accumulation and organization in computing and 39 communications industries: a regulationist approach', in P. Cooke, F. Moulaert, E. Swyngedouw, O. Weinstein and P. Wells (eds), Towards Global Localisation: The Computing and Telecommunications Industries in Britain and France, London: UCL Press, pp. 39-61.

Moulaert, F. and J. C. Delvainquiere (1994), 'Regional and sub-regional development in Europe: the role of socio-cultural trajectories', in L. Bekemans (ed.), Culture: Building stone for Europe 2002, Brussels: European University Press.

Moulaert, F. and L. Leontidou (1994), 'Localités désintégrées et strategies de la lutte contre la pauvreté: une réflexion méthodologique post-moderne', Espaces et sociétés, 78 (4), 35-54.

Moulaert, F. and A. J. Scott (eds) (1997), Cities, Enterprise and Society, A&C Black.

Moulaert, F. and P. M. Delladetsima (1998), Producer Services and Regional Development in the Aegean with Particular Focus on Lesvos and Chios, Brussels: Directorate General XII-F5.

Moulaert, F. and F. Sekia (2003), 'Territorial innovation models: a critical survey', Regional Studies, 37 (3), 289-302.

Moulaert, F. and J. Nussbaumer (2005), 'Defining the social economy and its governance at the neighbourhood level: a methodological reflection', Urban Studies, 42 (11), 2071-2088.

Moulaert, F. and K. Cabaret (2006), 'Planning, networks and power relations: Is democratic planning under capitalism possible?', Planning Theory, 5 (1), 51-70.

Moulaert, F. and A. Hamdouch (2006), 'New views of innovation systems: agents, rationales, networks and spatial scales in the knowledge infrastructure', Innovation: The European Journal of Social Science Research, 19 (1), 11-24.

Moulaert, F. and E. Christiaens (2010), 'The end of social innovation in urban development strategies? The case of BOM in Antwerp', in F. Moulaert, F. Martinelli, E. Swyngedouw, S. González (eds), Can Neighbourhoods Save the City? Community Development and Social Innovation, Oxford/New York: Routledge, pp. 168-184.

Moulaert, F. and A. Mehmood (2010), 'Analysing regional development: A structural realist approach', Regional Studies, 44 (1), 103-118.

Moulaert, F. and B. Jessop (2013), 'Theoretical foundations for the analysis of socio-economic development in space', in F. Martinelli, F. Moulaert and A. Novy (eds), Urban and Regional Development Trajectories in Contemporary Capitalism, London: Routledge.

Moulaert, F. and B. Van Dyck (2013), 'Framing social innovation research: a sociology of knowledge perspective', in F. Moulaert, D. MacCallum, A. Mehmood and A. Hamdouch (eds), The International Handbook on Social Innovation, Cheltenham:

Edward Elgar Publishing, pp. 466-479.

Moulaert, F. and A. Mehmood (2014), 'Towards social holism: Social innovation, holistic research methodology and pragmatic collective action', in E. Silva, P. Healey, N. Harris and P. Van den Broeck (eds), The Routledge Handbook of Planning Research Methods, London: Routledge, pp. 97-106.

Moulaert, F. and J. Nussbaumer (2014), 'Pour repenser l'innovation: vers un système régional d'innovation sociale', in J. L. Klein, J. L. Laville and F. Moulaert (eds), L'innovation sociale, Toulouse: Érès, pp. 81-113.

Moulaert, F. and A. Mehmood (2018), 'Towards a Social Innovation (SI) based epistemology in local development analysis: Lessons from twenty years of EU research', European Planning Studies, (in press).

Moulaert, F. and D. MacCallum (2019), Advanced Introduction to Social Innovation, Cheltenham: Edward Elgar Publishing.

Moulaert, F., E. Swyngedouw and P. Wilson (1988), 'Spatial responses to Fordist and post-Fordist accumulation and regulation', Papers in Regional Science, 64 (1), 11-23.

Moulaert, F., A. A. Aller, P. Cooke, C. Courlet, H. Häusserman and A. da Rosa (1990), Integrated area development and efficacy of local action. Feasibility study for the European Commission, Brussels: EC, DG Social Policy.

Moulaert, F., P. M. Delladetsima and L. Leontidou (eds) (1994), Local Development Strategies in Economically Disintegrated Areas: A Proactive Strategy Against Poverty in the European Community, Social Papers, Brussels: European Commission (DG V).

Moulaert, F., E. Salin and T. Werquin (2001a), 'Euralille: Large scale urban development and social polarization in the city', European Urban and Regional Studies, 8 (2), 145-160.

Moulaert, F., E. Swyngedouw and A. Rodriguez (2001b), 'Large scale urban development projects and local governance: from democratic urban planning to besieged local governance', Geographische Zeitschrift, 89 (H. 2/3), 71-84.

Moulaert, F., A. Rodriguez and E. Swyngedouw (eds) (2003), The Globalized City: Economic Restructuring and Social Polarization in the City, Oxford: Oxford University Press.

Moulaert, F., H. Demuynck and J. Nussbaumer (2004), 'Urban renaissance: from physical beautification to social empowerment: Lessons from Bruges—Cultural Capital of Europe 2002', City, 8 (2), 229-235.

Moulaert, F., F. Martinelli and E. Swyngedouw (eds) (2005), 'Social Innovation in the Governance of Urban Communities: A Multidisciplinary Perspective', special issue of Urban Studies, 42 (11).

Moulaert, F., F. Martinelli, S. González and E. Swyngedouw (eds) (2007a), 'Social Innovation and Governance in European Cities', special issue of European Urban and Regional Studies, 14 (3).

Moulaert, F., F. Martinelli, S, González and E. Swyngedouw (2007b), 'Social innovation and governance in European cities: urban development between path dependency and radical innovation', European Urban and Regional Studies, 14 (3), 195-209.

Moulaert, F., E. Morlicchio and L. Cavola (2007c), 'Social exclusion and urban policy: Combining "Northern" and "Southern European" perspectives', in H. S. Geyer (ed.), International Handbook of Urban Policy, Cheltenham: Edward Elgar Publishing, Volume 1, pp. 138-158.

Moulaert, F., F. Martinelli, E. Swyngedouw and S. González (eds) (2010), Can Neighbourhoods Save the City? Community Development and Social Innovation, London: Routledge.

Moulaert, F., D. MacCallum and J. Hillier (2013a), 'Social innovation: Intuition, precept, concept, theory and practice', in F. Moulaert, D. MacCallum, A. Mehmood and A. Hamdouch (eds), The International Handbook on Social Innovation: Collective Action, Social Learning and Transdisciplinary Research, London: Edward Elgar Publishing, pp. 13-25.

Moulaert, F., D. MacCallum, A. Mehmood and A. Hamdouch (eds) (2013b), The International Handbook on Social Innovation: Collective Action, Social Learning and Transdisciplinary Research, Cheltenham: Edward Elgar Publishing.

Moulaert, F., B. Van Dyck, A. Z. Khan and J. Schreurs (2013c), 'Building a meta-framework to 'address' spatial quality', International Planning Studies, 18 (3-4), 389-409.

Moulaert, F. with P. Caballero, M. Macharia, M. Midheme and E. Wamuchiru (2014), Housing at the time of the Third Enclosure, lecture at the Hollands, 6 November.

Moulaert, F., B. Jessop and A. Mehmood (2016a), 'Agency, structure, institutions, discourse (ASID) in urban and regional development', International Journal of Urban Sciences, 20 (2), 167-187.

Moulaert, F., B. Van Dyck and C. Parra (2016b), 'Social science's say on the governance of socio-ecological development: when rational agents become human', Workshop on Ecology and Social Sciences, Leuven Space and Society, 12 January 2016, Leuven.

Moulaert, F., A. Mehmood, D. MacCallum and B. Leubolt (2017), Social innovation as a trigger for transformations – the role of research, Luxembourg: Publications Office of the European Union.

Mulgan, G., S. Tucker, R. Ali and B. Sanders (2007), Social Innovation: What it is, Why it Matters and How It Can be Accelerated, London: The Young Foundation.

Munteanu, M. and L. Servillo (2014), 'The Romanian planning system: Post-communist dynamics of change and Europeanization processes', European Planning Studies, 22 (11), 2248-2267.

Murray, R., J. Caulier-Grice and G. Mulgan (2010), The Open Book of Social Innovation. London: NESTA/Young Foundation.

Nelson, R. R. (1987), 'Innovation and economic development Theoretical reprospect and prospect', in J. M. Katz (ed.), Technology Generation in Latin American Manufacturing Industries, London: Palgrave Macmillan, pp. 78-93.

Nicholls, A., J. Simon and M. Gabriel (eds) (2015), New Frontiers in Social Innovation Research, London: Palgrave Macmillan.

Nichols, N., D. J. Phipps, J. Provençal and A. Hewitt (2013), 'Knowledge mobilization, collaboration, and social innovation: Leveraging investments in higher education', Canadian Journal of Nonprofit and Social Economy Research, 4 (1), 25-42.

Novy, A. (2012), '"Unequal diversity" as a knowledge alliance: An encounter of Paulo Freire's dialogical approach and transdisciplinarity', Multicultural Education & Technology Journal, 6 (3), 137-148.

Novy, A., D. C. Swiatek and F. Moulaert (2012), 'Social cohesion: a conceptual and political elucidation', Urban Studies, 49 (9), 1873-1889.

Novy, A., S. Habersack and B. Schaller (2013), 'Innovative forms of knowledge production: Transdisciplinarity and knowledge alliances', in F. Moulaert, D. MacCallum, A. Mehmood and A. Hamdouch (eds), The International Handbook of Social Innovation. Collective Action, Social Learning and Transdisciplinary Research, Cheltenham: Edward Elgar Publishing, pp. 442-452.

Novy, A. (2014), Shaping the Great Transformation: Implications for Europe, Vienna: Grüne Bildungswerkstatt.

OECD (2010), SMEs, Entrepreneurship and Innovation, Paris: OECD.

Oosterlynck, S., A. Novy, Y. Kazepov, P. Cools, T. Saruis, B. Leubolt and F. Wukovitsch (forthcoming), 'Improving poverty reduction: lessons from the social innovation perspective', in B. Cantillon, T. Goedomé and J. Hills (eds), Improving Poverty Reduction in Europe. Lessons from the past, scenarios for the future, Oxford: Oxford University Press.

Paidakaki, A. (2017), Uncovering the Housing-Resilience Nexus. Social Resilience Cells, Governance Hybridities and (Hetero-)Production of Post-Disaster 'Egalitarian' Urbanities. The Case Study of Post-Katrina New Orleans (Doctoral dissertation, KU Leuven).

Paidakaki, A. and F. Moulaert (2017), 'Does the post-disaster resilient city really exist? A critical analysis of the heterogeneous transformative capacities of housing reconstruction "resilience cells"', International Journal of Disaster Resilience in the Built Environment, 8 (3), 275-291.

Paidakaki, A. and F. Moulaert (2018), 'Disaster resilience into which direction(s)? Competing discursive and material practices in post-Katrina New Orleans', Housing, Theory and Society, 35(4), 432-454.

Parra, C. (2010), The governance of ecotourism as a socially innovative force for paving the way for more sustainable development paths: the Morvan regional park (Doctoral dissertation, Université des Sciences et Technologies de Lille).

Parra, C. (2013), 'Social sustainability: a competing concept to social innovation?', in F. Moulaert, D. MacCallum, A. Mehmood and A. Hamdouch (eds), The International Handbook of Social Innovation, Cheltenham: Edward Elgar Publishing, pp. 142-154.

Parra, C. and F. Moulaert (2010), 'Why sustainability is so fragilely 'social'...', in S. Oosterlynck, J. Van den Broeck, L. Albrechts, F. Moulaert and A. Verhetsel (eds), Strategic Spatial Projects: Catalysts for Change, London: Routledge, pp. 163-173.

Parra, C. and F. Moulaert (2011), La nature de la durabilité sociale: vers une lecture socioculturelle du développement territorial durable, Développement durable et territoires. Économie, géographie, politique, droit, sociologie, 2 (2).

Parra, C. and F. Moulaert (2016), 'The Governance of the nature-culture nexus: Lessons learned from the San Pedro de Atacama case study', Nature and Culture, 11 (3),

239-258.

Pearson, J. (2008), Going Local. Working in communities and neighborhoods, London: Routledge.

Perera, N. (2016), People's Spaces. Coping, Familiarising, Creating, New York/London: Routledge.

Perrin, N. and Q. Schoonvaere (2008), Migrations et populations issues de l'immigration en Belgique. Rapport statistique et démographique 2008 (Migrations and immigrant populations in Belgium. Demographic and statistical report 2008), Brussels: Centre pour l'égalité des chances et la lutte contre le racisme (Centre for Equal Opportunities and Opposition to Racism), accessed 2016 at www. diversité.be; and now at https://www.unia.be/en.

Pinder, D. (2015), 'Reconstituting the possible: Lefebvre, utopia and the urban question', International Journal of Urban and Regional Research, 39 (1), 28-45.

Polanyi, K. (1944), The Great Transformation: The Political and Economic Origins of Our Times, New York: Farrar and Rinehart.

Polanyi, K. (2001), The Great Transformation: The Political and Economic Origins of our Time, Boston: Beacon Press (Originally published: New York: Farrar and Rinehart, 1944).

POLEKAR (ed.) (1981), Krisis en Werkgelegenheid, Leuven: Kritak.

POLEKAR (ed.) (1985), Het Laboratorium van de Crisis, Leuven: Kritak.

Pradel, M., M. Garcia and S. Eizaguirre (2013), 'Theorizing multi-level governance in social innovation dynamics', in F. Moulaert, D. MacCallum, A. Mehmood and A. Hamdouch (eds), The International Handbook of Social Innovation, Cheltenham: Edward Elgar Publishing, pp. 155-168.

Pratschke, J. and E. Morlicchio (2012), 'Social polarization, the labour market and economic restructuring in Europe: An urban perspective', Urban Studies, 49 (9), 1891-1907.

Putri, P. W. (2014), Black Water-Grey Settlements. Domestic Wastewater Management and the Socio-ecological Dynamics of Jakarta's Kampungs (Doctoral dissertation, KU Leuven).

Rodriguez, A. and E. Martinez, (2003), 'Restructuring cities: miracles and mirages in urban revitalization in Bilbao', in Moulaert, F., A. Rodriguez and E. Swyngedouw (eds) (2003), The Globalized City: Economic Restructuring and Social Polarization in the City, Oxford: Oxford University Press, pp. 181-207.

Ruby, C. (2002), 'L'art public dans la ville', accessed 2 March 2011 at http://www. espacestemps.net /articles/art-public-dans-la-ville/.

Sack, R. D. (1992), Place, Modernity and the Consumer's World, Baltimore: Johns Hopkins.

Santos, B. D. S. (2011), 'Épistémologies du Sud', Études rurales, (187), 21-49.

Sayer, A. (1992), Method in Social Science: A Realist Approach (2 ed.), London: Routledge.

Segers, R., P. Van den Broeck, A. Khan, F. Moulaert, J. Schreurs, B. De Meulder, K. Miciukiewicz, G. Vigar and A. Madanipour (eds) (2016), The SPINDUS handbook for spatial quality. A relational approach, Brussels: Academic and Scientific Publishers (ASP).

Sen, A. (2001), Development as Freedom, Oxford: Oxford University Press.

Servillo, L. (2017), 'Strategic planning and institutional change, a karst river phenomenon', in L. Albrechts, A. Balducci and J. Hillier (eds), Situated Practices of Strategic Planning. London: Routledge, pp. 331-347.

Servillo, L. and V. Lingua (2014), 'The innovation of the Italian planning system: Actors, path dependencies, cultural contradictions and a missing epilogue', European Planning Studies, 22 (2), 400-417.

Servillo, L. and P. Van den Broeck (2012), 'The social construction of planning systems. A strategic-relational institutionalist approach', Planning Practice & Research, 27 (1), 41-61.

Smiers, J. (2003), Arts under Pressure: Protecting cultural diversity in the age of globalization, Zed Books.

Smith, A. and A. Stirling (2010), 'The politics of social-ecological resilience and sustainable socio-technical transitions', Ecology and Society, 15 (1), 11.

Smith, A., A. Ely, M. Fressoli, D. Abrol and E. Arond (2016), Grassroots Innovation Movements, Routledge.

Söderbaum, P. (2000), Ecological Economics: A political economics approach to environment and development. Earthscan.

Stern, M. J. and S. C. Seifert (2010), 'Cultural clusters: The implications of cultural assets agglomeration for neighborhood revitalization', Journal of Planning Education and Research, 29 (3), 262-279.

Sternberg, E. (2000), 'An integrative theory of urban design', Journal of the American Planning Association, 66 (3), 265-278.

Stohr, W. B. and F. F. Todtling (1978), 'Spatial enquiry-some antitheses to current regional development strategy'. Regional Science Association, 38.

Straw, W. (2004), 'Cultural scenes', Loisir et société/Society and Leisure, 27 (2), 411-422.

Sum, N. L. and B. Jessop (2013), Towards a Cultural Political Economy: Putting Culture in its Place in Political Economy, Cheltenham: Edward Elgar Publishing.

Surikova, S., K. Oganisjana and G. Grinberga-Zalite (eds) (2015), 'The role of education in promoting social innovation processes in the society', Rēzeknes Augstskola, 4, 234-243.

Swyngedouw, E. (2005), 'Governance innovation and the citizen: the Janus face of governance-beyond-the-state', Urban Studies, 42 (11), 1991-2006.

Swyngedouw, E. and F. Moulaert (2010), 'Socially innovative projects, governance dynamics and urban change: between state and self-organization', in F. Moulaert, F. Martinelli, E. Swyngedouw and S. González (eds), Can Neighbourhoods Save the City? Community Development and Social Innovation, London: Routledge, pp. 219-234.

Swyngedouw, E., F. Moulaert and A. Rodriguez (2002), 'Neoliberal urbanisation in Europe: large-scale urban development projects and the new urban policy', Antipode, 34 (3), 542-577.

Tickell, A. and J. Peck (1992), 'Accumulation, regulation and the geographies of post-Fordism: missing links in regulationist research', Progress in Human Geography, 16 (2), 190-218.

Tuan, Y. F. (1977), Space and place: The perspective of experience, University of Minnesota Press.

Unger, R. M. (2015), 'Conclusion: the task of the social innovation movement', in A. Nicolls, J. Simon and M. Gabriel (eds), New Frontiers in Social Innovation Research, Basingstoke: Palgrave, pp. 233-251.

Vahtrapuu, A. (2013), 'Le rôle des artistes dans la revitalisation des espaces urbains en déclin: pour une approche sensorielle de la ville', Territoire en mouvement Revue de géographie et aménagement. Territory in movement Journal of Geography and Planning, (17-18), 103-116.

Van den Broeck, P. (2011), 'Analysing social innovation through planning instruments. A strategic-relational approach.', in S. Oosterlynck, J. Van den Broeck, L. Albrechts, F. Moulaert and A. Verhetsel (eds), Strategic projects. Catalysts for change, London/New York: Routledge, pp. 52-78.

Van den Broeck, P. and K. Verachtert (2016), 'Whose permits? The tenacity of permissive development control in Flanders', European Planning Studies, 24 (2), 387-406.

Van den Broeck, P., F. Moulaert, A. Kuhk, E. Lievois and J. Schreurs (2014), 'Spatial planning in Flanders. Serving a by-passed capitalism?', in H. Blotevogel, P. Getimis and M. Reimer (eds), Spatial Planning Systems and Practices in Europe: Towards Multiple Trajectories of Change, London: Routledge, pp. 190-209.

Van den Broeck, P., B. Leubolt, F. Moulaert, B. Hubeau, P. Delladetsimas, G. Vloebergh and A. Kuhk (forthcoming), Understanding land use rights, building landed commons, INDIGO research project.

Van Dyck, B. and P. Van den Broeck (2013), 'Social innovation: a territorial process', in F. Moulaert, D. MacCallum, A. Mehmood and A. Hamdouch (eds), The International Handbook on Social Innovation. Collective Action, Social Learning and Transdisciplinary Research, Cheltenham: Edward Elgar Publishing, pp. 131-141.

Van Hove, E. (2001), Networking Neighbourhoods, Columbia, SC: University of South Carolina Press.

Wagenaar, H. and J. Van der Heijden (2015), 'The promise of democracy? Civic enterprise, localism and the transformation of democratic capitalism', in S. Davoudi and A. Madanipour (eds), Re-Considering Localism, London: Routledge, pp. 126-146.

Wagenaar, H., P. Healey, G. Laino, P. Healey, G. Vigar, S. Riutort Isern, … and H. Wagenaar (2015), 'The transformative potential of civic enterprise', Planning Theory & Practice, 16 (4), 557-585.

Wenger, E. (1998), Communities of Practice: Learning, meaning and identity, Cambridge: Cambridge University Press.

Western Australia (WA) Department of Planning, n.d. 'Aboriginal Settlements', accessed 15 May 2016 at https://www.planning.wa.gov.au/Aboriginal-settlements.asp.

Williams, A., M. Goodwin and P. Cloke (2014), 'Neoliberalism, big society, and progressive localism', Environment & Planning A, 46 (12), 2798-2815.

Wilson, J. and E. Swyngedouw (eds) (2014), The Post-Political and its Discontents, Edinburgh: Edinburgh University Press.

Young Foundation (2010), The Open Book of Social Innovation, London: Young Foundation, NESTA.

INDEX